Yamaha RD250 & 350LC Owners Workshop Manual

by Pete Shoemark

Models covered
RD250 LC. 247cc. Introduced May 1980
RD350 LC. 347cc. Introduced June 1980

ISBN 978 0 85696 803 7

© Haynes Group Limited 1993

All rights reserved. No part of this book may be reproduced or transmitted in any form or by any means, electronic or mechanical, including photocopying, recording or by any information storage or retrieval system, without permission in writing from the copyright holder.

(803-7S7)

Haynes Group Limited
Haynes North America, Inc

www.haynes.com

Acknowledgements

Our thanks are due to Fran Ridewood and Co of Wells, who supplied the RD250 LC featured throughout this manual and to Mitsui Machinery Sales (UK) Ltd who supplied the necessary service information and gave permission to reproduce many of the line drawings used.

The Avon Rubber Company supplied information and technical assistance on tyre care and fitting, and NGK Spark Plugs (UK) Ltd provided information on plug maintenance and electrode conditions. Renold Ltd advised on chain maintenance and renewal.

About this manual

The purpose of this manual is to present the owner with a concise and graphic guide which will enable him to tackle any operation from basic routine maintenance to a major overhaul. It has been assumed that any work will be undertaken without the luxury of a well-equipped workshop and a range of manufacturer's service tools.

To this end, the machine featured in the manual was stripped and rebuilt in our own workshop, by a team comprising a mechanic, a photographer and the author. The resulting photographic sequence depicts events as they took place, the hands shown being those of the author and the mechanic.

The use of specialised, and expensive, service tools was avoided unless their use was considered to be essential due to risk of breakage or injury. There is usually some way of improvising a method of removing a stubborn component, provided that a suitable degree of care is exercised.

The author learnt his motorcycle mechanics over a number of years, faced with the same difficulties and using similar facilities to those encountered by most owners. It is hoped that this practical experience can be passed on through the pages of this manual.

Where possible, a well-used example of the machine is chosen for a workshop project, as this highlights any areas which might be particularly prone to giving rise to problems. In this way, any such difficulties are encountered and resolved before the text is written, and the techniques used to deal with them can be incorporated in the relevant Section. Armed with a working knowledge of the machine, the author undertakes a considerable amount of research in order that the maximum amount of data can be included in the manual.

Each Chapter is divided into numbered sections. Within these Sections are numbered paragraphs. Cross reference throughout the manual is quite straightforward and logical. When reference is made 'See Section 6.10' it means Section 6, paragraph 10 in the same Chapter. If another Chapter were intended, the reference would read, for example, 'See Section 2, Section 6.10'. All the photographs are captioned with a section/paragraph number to which they refer and are relevant to the Chapter text adjacent.

Figures (usually line illustrations) appear in a logical but numerical order, within a given Chapter. Fig 1.1 therefore refers to the first figure in Chpater 1.

Left-hand and right-hand descriptions of the machines and their components refer to the left and right of a given machine when the rider is seated normally.

Motorcycle manufacturers continually make changes to specifications and recommendations and these, when notified, are incorporated into our manuals at the earliest opportunity.

We take great pride in the accuracy of information given in this manual, but motorcycle manufacturers make alterations and design changes during the production run of a particular motorcycle of which they do not inform us. No liability can be accepted by the authors or publishers for loss, damage or injury caused by any errors in, or omissions from, the information given.

Contents

Chapter	Section	Page	Section	Page
Preliminary sections	Acknowledgements	2	Ordering spare parts	7
	About this manual	2	Safety first!	8
	Introduction to the Yamaha		Working conditions and tools	9
	RD250 and RD350 LC		Fault diagnosis	10-19
	models	6	Routine maintenance	20-31
	Dimensions and weights	6	Castrol lubricants	32
Chapter 1 Engine, clutch and gearbox	Operations with unit in frame	35	Gearbox components	60
			Clutch	63
	Unit dismantling – general	40	Engine reassembly – general	63
	Examination and renovation:		Starting and running the	
	General	47	rebuilt engine	85
	Engine	58		
Chapter 2 Cooling system	Draining the cooling system	88	Water pump	92
	Radiator and cap	89	Water temperature gauge	92
	Hoses	91		
Chapter 3 Fuel system and lubrication	Petrol tank	96	Air cleaner	105
	Petrol cap	96	Lubrication – general	
	Carburettors	97,98,102,103	description	105
	Exhaust system	103	Oil pump	107
			Reed valve induction system	110
Chapter 4 Ignition system	Testing electronic ignition system	114	Coils – testing	116
			Ignition timing	117
	Wiring – checking	114	Spark plugs	117
Chapter 5 Frame and forks	Front forks	119,123	Rear suspension assembly	128,133
	Steering head assembly	121	Rear sub frame	128
	Steering head bearings	126	Centre stand	134
	Steering lock	126	Prop stand	134
	Frame – examination and		Footrests	134
	renovation	127	Speedometer and tachometer	134
Chapter 6 Wheels, brakes and tyres	Front wheel	137	Rear wheel bearings	149
	Disc brake	137	Cush drive	153
	Front disc brake master		Rear wheel sprocket	154
	cylinder	140	Final drive chain	156
	Bleeding brakes	143	Tyres	156
	Rear wheel	146	Wheel balancing	157
	Front wheel bearings	149		
Chapter 7 Electrical system	Charging system	159	Headlamp	162
	Battery	159	Flashing indicator circuit	165
	Alternator – resistance tests	160	Instrument head	166
	Regulator/rectifier unit	160	Switches	169,170
	Fuses	161	Wiring diagram	172

Note: General description and specifications are given in each Chapter immediately after the list of Contents

Index   173 – 176

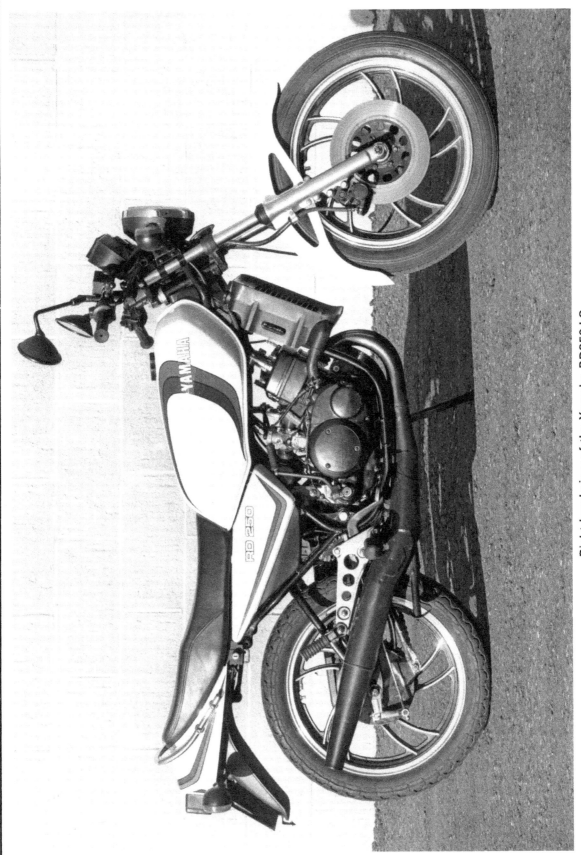

Right-hand view of the Yamaha RD250 LC

Engine/gearbox unit of the Yamaha RD250 LC

Introduction to the Yamaha RD250 and 350 LC models

The Yamaha RD250 LC and RD350 LC models represent what many feel will be the last of the high performance two-strokes. In recent years the increasing emphasis on reduced exhaust emissions and improved fuel consumption has occasioned a change of approach from those manufacturers whose previous speciality was the small, fast two-stroke machine. In the midst of a bewildering range of four-stroke machines of ever increasing sophistication, the LCs' arrival was something of a surprise.

The two models are very similar in all respects with the obvious exception of a few specification differences and the two capacities. In general terms, the LC models are similar in performance to the air-cooled RD models which they replace. The assumption that a machine which bore rather more than a passing resemblance to the sucessful Yamaha TZ racing machines would prove popular seems to have been proved correct.

It is interesting to note that the air-cooled RD models and the TZ racers shared a common ancestor in the YDS2 of the late 1960s. In the intervening years the two have gone their separate ways and developed into distinctly different machines. In recent years the RD models have come to be considered just a little old fashioned despite their being willing performers on the road and in production racing. What better than to model a new version of the RDs on the consistently successful TZ racers?

In fact, the LC models have far more in common with the air-cooled RD models than with the racing machines, but the attractions of TZ styling complete with water cooling and cantilever rear suspension have done much to attract owners with sporting inclinations. Like the earlier air-cooled models, the LCs have become firmly established in Production Racing.

The adoption of water cooling of the cylinder head and barrels means that the engine can be run at much closer tolerances than would be possible with a conventional air-cooled engine. A mixture of equal parts of water and corrosion inhibited anti-freeze is circulated around the water passages by a crankshaft-driven pump. The coolant is able to take heat away from any particular hot spots before distortion becomes evident, and helps to maintain an even running temperature in the engine. The hot coolant is then passesd through the frame-mounted aluminium radiator, and is thus cooled by airflow through its matrix. The system is not equipped with a thermostat or a cooling fan.

The engine is otherwise similar to the air-cooled RD models, though with many detail changes. Induction is controlled by reed valves to prevent blow back through the carburettors and thus allow less conservative port timing. Lubrication is by pump fed injection.

The chassis is of orthodox design, but incorporates the recently revived cantilever rear suspension arrangement in which the swinging arm fork is replaced by a welded subframe. A single central suspension unit is mounted between the cantilever and the frame backbone. The suspension unit is of the De Carbon type and incorporates a sealed nitrogen-filled section which is deparated from the damping oil by a floating piston. The nitrogen is compressed as the unit operates and is designed to supplement the main coil spring.

Dimensions and weight

	RD250LC	RD350LC
Overall length	2080 mm (81.89 in)	2080 mm (81.89 in)
Overall width	750 mm (29.53 in)	750 mm (29.53 in)
Overall height	1090 mm (42.91 in)	1090 mm (42.91 in)
Seat height	785 mm (30.90 in)	785 mm (30.90 in)
Wheelbase	1360 mm (53.54 mm)	1365 mm (53.74 in)
Ground clearance	165 mm (6.49 in)	165 mm (6.49 in)
Net weight	139 kg (306.5 lb)	143 kg (315.3 lb)

Ordering spare parts

Before attempting any overhaul or maintenance work it is important to ensure that any parts likely to be required are to hand. Many of the more common parts such as gaskets and seals will be available off the shelf form the local Yamaha dealer, but often it will prove necessary to order more specialised parts well in advance. It is worthwhile running through the operation to be undertaken, referring to the appropriate Chapter and section of this book, so that a note can be made of the items most likely to be required. In some instances it will of course be necessary to dismantle the assembly in question so that the various components can be examined and measured for wear and in these instances, it must be remembered that the machine may have to be left dismantled while the replacement parts are obtained.

It is advisable to purchase almost all new parts from an official Yamaha dealer. Almost any motorcycle dealer should be able to obtain the parts in time, but this may take longer than it would through the official factory spares arrangement. It is quite in order to purchase expendable items such as spark plugs, bulbs, tyres, oil and grease from the nearest convenient source.

Owners should be very wary of some of the pattern parts that might be offered at a lower price than the Yamaha originals. Whilst in most cases these will be of an adequate standard, some of the more important parts have been known to fail suddenly and cause extensive damage in the process. A particular danger in recent years is the growing numbers of counterfeit parts from Taiwan. These include items such as oil

filters and brake pads and are often sold in packaging which is almost indistinguishable from the manufacturer's own. Again, these are often quite serviceable parts, but can sometimes be dangerously inadequate in materials or construction. Apart from rendering the manufacturer's warranty invalid, use of substandard parts may put the life of the rider (or the machine) at risk. In short, where there are any doubts on safety grounds purchase parts **only** from a reputable Yamaha dealer. The extra cost involved pays for a high standard of quality and the parts will be guaranteed to work effectively.

Most machines are subject to continuous detail modifications throughout their production run, and in addition to annual model changes. In most cases these changes will be known to the dealer but not to the general public, so it is essential to quote the engine and frame numbers in full when ordering parts. The engine number is embossed in a rectangular section of the crankcase just below the left-hand carburettor, and the frame number is stamped on the right-hand side of the steering head.

Owners of the LC models may wish to modify their machines for road or racing use, but before doing so remember that the machine's warranty may be affected. Always choose accessories from a reputable manufacturer. If a modified exhaust system is being considered make sure that it is of good design and construction and that it has been shown to improve performance. A reputable manufacturer will provide information on any carburettor jet changes required to suit his particular system.

Location of engine number

Location of frame number

Safety first!

Professional motor mechanics are trained in safe working procedures. However enthusiastic you may be about getting on with the job in hand, do take the time to ensure that your safety is not put at risk. A moment's lack of attention can result in an accident, as can failure to observe certain elementary precautions.

There will always be new ways of having accidents, and the following points do not pretend to be a comprehensive list of all dangers; they are intended rather to make you aware of the risks and to encourage a safety-conscious approach to all work you carry out on your vehicle.

Essential DOs and DON'Ts

DON'T start the engine without first ascertaining that the transmission is in neutral.

DON'T suddenly remove the filler cap from a hot cooling system – cover it with a cloth and release the pressure gradually first, or you may get scalded by escaping coolant.

DON'T attempt to drain oil until you are sure it has cooled sufficiently to avoid scalding you.

DON'T grasp any part of the engine, exhaust or silencer without first ascertaining that it is sufficiently cool to avoid burning you.

DON'T allow brake fluid or antifreeze to contact the machine's paintwork or plastic components.

DON'T syphon toxic liquids such as fuel, brake fluid or antifreeze by mouth, or allow them to remain on your skin.

DON'T inhale dust – it may be injurious to health (see *Asbestos* heading).

DON'T allow any spilt oil or grease to remain on the floor – wipe it up straight away, before someone slips on it.

DON'T use ill-fitting spanners or other tools which may slip and cause injury.

DON'T attempt to lift a heavy component which may be beyond your capability – get assistance.

DON'T rush to finish a job, or take unverified short cuts.

DON'T allow children or animals in or around an unattended vehicle.

DON'T inflate a tyre to a pressure above the recommended maximum. Apart from overstressing the carcase and wheel rim, in extreme cases the tyre may blow off forcibly.

DO ensure that the machine is supported securely at all times. This is especially important when the machine is blocked up to aid wheel or fork removal.

DO take care when attempting to slacken a stubborn nut or bolt. It is generally better to pull on a spanner, rather than push, so that if slippage occurs you fall away from the machine rather than on to it.

DO wear eye protection when using power tools such as drill, sander, bench grinder etc.

DO use a barrier cream on your hands prior to undertaking dirty jobs – it will protect your skin from infection as well as making the dirt easier to remove afterwards; but make sure your hands aren't left slippery. Note that long-term contact with used engine oil can be a health hazard.

DO keep loose clothing (cuffs, tie etc) and long hair well out of the way of moving mechanical parts.

DO remove rings, wristwatch etc, before working on the vehicle – especially the electrical system.

DO keep your work area tidy – it is only too easy to fall over articles left lying around.

DO exercise caution when compressing springs for removal or installation. Ensure that the tension is applied and released in a controlled manner, using suitable tools which preclude the possibility of the spring escaping violently.

DO ensure that any lifting tackle used has a safe working load rating adequate for the job.

DO get someone to check periodically that all is well, when working alone on the vehicle.

DO carry out work in a logical sequence and check that everything is correctly assembled and tightened afterwards.

DO remember that your vehicle's safety affects that of yourself and others. If in doubt on any point, get specialist advice.

IF, in spite of following these precautions, you are unfortunate enough to injure yourself, seek medical attention as soon as possible.

Asbestos

Certain friction, insulating, sealing, and other products – such as brake linings, clutch linings, gaskets, etc – contain asbestos. *Extreme care must be taken to avoid inhalation of dust from such products since it is hazardous to health.* If in doubt, assume that they *do* contain asbestos.

Fire

Remember at all times that petrol (gasoline) is highly flammable. Never smoke, or have any kind of naked flame around, when working on the vehicle. But the risk does not end there – a spark caused by an electrical short-circuit, by two metal surfaces contacting each other, by careless use of tools, or even by static electricity built up in your body under certain conditions, can ignite petrol vapour, which in a confined space is highly explosive.

Always disconnect the battery earth (ground) terminal before working on any part of the fuel or electrical system, and never risk spilling fuel on to a hot engine or exhaust.

It is recommended that a fire extinguisher of a type suitable for fuel and electrical fires is kept handy in the garage or workplace at all times. Never try to extinguish a fuel or electrical fire with water.

Note: *Any reference to a 'torch' appearing in this manual should always be taken to mean a hand-held battery-operated electric lamp or flashlight. It does **not** mean a welding/gas torch or blowlamp.*

Fumes

Certain fumes are highly toxic and can quickly cause unconsciousness and even death if inhaled to any extent. Petrol (gasoline) vapour comes into this category, as do the vapours from certain solvents such as trichloroethylene. Any draining or pouring of such volatile fluids should be done in a well ventilated area.

When using cleaning fluids and solvents, read the instructions carefully. Never use materials from unmarked containers – they may give off poisonous vapours.

Never run the engine of a motor vehicle in an enclosed space such as a garage. Exhaust fumes contain carbon monoxide which is extremely poisonous; if you need to run the engine, always do so in the open air or at least have the rear of the vehicle outside the workplace.

The battery

Never cause a spark, or allow a naked light, near the vehicle's battery. It will normally be giving off a certain amount of hydrogen gas, which is highly explosive.

Always disconnect the battery earth (ground) terminal before working on the fuel or electrical systems.

If possible, loosen the filler plugs or cover when charging the battery from an external source. Do not charge at an excessive rate or the battery may burst.

Take care when topping up and when carrying the battery. The acid electrolyte, even when diluted, is very corrosive and should not be allowed to contact the eyes or skin.

If you ever need to prepare electrolyte yourself, always add the acid slowly to the water, and never the other way round. Protect against splashes by wearing rubber gloves and goggles.

Mains electricity and electrical equipment

When using an electric power tool, inspection light etc, always ensure that the appliance is correctly connected to its plug and that, where necessary, it is properly earthed (grounded). Do not use such appliances in damp conditions and, again, beware of creating a spark or applying excessive heat in the vicinity of fuel or fuel vapour. Also ensure that the appliances meet the relevant national safety standards.

Ignition HT voltage

A severe electric shock can result from touching certain parts of the ignition system, such as the HT leads, when the engine is running or being cranked, particularly if components are damp or the insulation is defective. Where an electronic ignition system is fitted, the HT voltage is much higher and could prove fatal.

Working conditions and tools

When a major overhaul is contemplated, it is important that a clean, well-lit working space is available, equipped with a workbench and vice, and with space for laying out or storing the dismantled assemblies in an orderly manner where they are unlikely to be disturbed. The use of a good workshop will give the satisfaction of work done in comfort and without haste, where there is little chance of the machine being dismantled and reassembled in anything other than clean surroundings. Unfortunately, these ideal working conditions are not always practicable and under these latter circumstances when improvisation is called for, extra care and time will be needed.

The other essential requirement is a comprehensive set of good quality tools. Quality is of prime importance since cheap tools will prove expensive in the long run if they slip or break when in use, causing personal injury or expensive damage to the component being worked on. A good quality tool will last a long time, and more than justify the cost.

For practically all tools, a tool factor is the best source since he will have a very comprehensive range compared with the average garage or accessory shop. Having said that, accessory shops often offer excellent quality tools at discount prices, so it pays to shop around. There are plenty of tools around at reasonable prices, but always aim to purchase items which meet the relevant national safety standards. If in doubt, seek the advice of the shop proprietor or manager before making a purchase.

The basis of any tool kit is a set of open-ended spanners, which can be used on almost any part of the machine to which there is reasonable access. A set of ring spanners makes a useful addition, since they can be used on nuts that are very tight or where access is restricted. Where the cost has to be kept within reasonable bounds, a compromise can be effected with a set of combination spanners – open-ended at one end and having a ring of the same size on the other end. Socket spanners may also be considered a good investment, a basic 3/8 in or 1/2 in drive kit comprising a ratchet handle and a small number of socket heads, if money is limited. Additional sockets can be purchased, as and when they are required. Provided they are slim in profile, sockets will reach nuts or bolts that are deeply recessed. When purchasing spanners of any kind, make sure the correct size standard is purchased. Almost all machines manufactured outside the UK and the USA have metric nuts and bolts, whilst those produced in Britain have BSF or BSW sizes. The standard used in USA is AF, which is also found on some of the later British machines. Others tools that should be included in the kit are a range of crosshead screwdrivers, a pair of pliers and a hammer.

When considering the purchase of tools, it should be remembered that by carrying out the work oneself, a large proportion of the normal repair cost, made up by labour charges, will be saved. The economy made on even a minor overhaul will go a long way towards the improvement of a toolkit.

In addition to the basic tool kit, certain additional tools can prove invaluable when they are close to hand, to help speed up a multitude of repetitive jobs. For example, an impact screwdriver will ease the removal of screws that have been tightened by a similar tool, during assembly, without a risk of damaging the screw heads. And, of course, it can be used again to retighten the screws, to ensure an oil or airtight seal results. Circlip pliers have their uses too, since gear pinions, shafts and similar components are frequently retained by circlips that are not too easily displaced by a screwdriver. There are two types of circlip pliers, one for internal and one for external circlips. They may also have straight or right-angled jaws.

One of the most useful of all tools is the torque wrench, a form of spanner that can be adjusted to slip when a measured amount of force is applied to any bolt or nut. Torque wrench settings are given in almost every modern workshop or service manual, where the extent to which a complex component, such as a cylinder head, can be tightened without fear of distortion or leakage. The tightening of bearing caps is yet another example. Overtightening will stretch or even break bolts, necessitating extra work to extract the broken portions.

As may be expected, the more sophisticated the machine, the greater is the number of tools likely to be required if it is to be kept in first class condition by the home mechanic. Unfortunately there are certain jobs which cannot be accomplished successfully without the correct equipment and although there is invariably a specialist who will undertake the work for a fee, the home mechanic will have to dig more deeply in his pocket for the purchase of similar equipment if he does not wish to employ the services of others. Here a word of caution is necessary, since some of these jobs are best left to the expert. Although an electrical multimeter of the AVO type will prove helpful in tracing electrical faults, in inexperienced hands it may irrevocably damage some of the electrical components if a test current is passed through them in the wrong direction. This can apply to the synchronisation of twin or multiple carburettors too, where a certain amount of expertise is needed when setting them up with vacuum gauges. These are, however, exceptions. Some instruments, such as a strobe lamp, are virtually essential when checking the timing of a machine powered by CDI ignition system. In short, do not purchase any of these special items unless you have the experience to use them correctly.

Although this manual shows how components can be removed and replaced without the use of special service tools (unless absolutely essential), it is worthwhile giving consideration to the purchase of the more commonly used tools if the machine is regarded as a long term purchase Whilst the alternative methods suggested will remove and replace parts without risk of damage, the use of the special tools recommended and sold by the manufacturer will invariably save time.

Fault diagnosis

Contents

Introduction 1

Engine does not start when turned over
No fuel flow to carburettor 2
Fuel not reaching cylinder 3
Engine flooding 4
No spark at plug 5
Weak spark at plug 6
Compression low 7

Engine stalls after starting
General causes 8

Poor running at idle and low speed
Weak spark at plug or erratic firing 9
Fuel/air mixture incorrect 10
Compression low 11

Acceleration poor
General causes 12

Poor running or lack of power at high speeds
Weak spark at plug or erratic firing 13
Fuel/air mixture incorrect 14
Compression low 15

Knocking or pinking
General causes 16

Overheating
Firing incorrect 17
Fuel/air mixture incorrect 18
Lubrication inadequate 19
Miscellaneous causes 20

Clutch operating problems
Clutch slip 21
Clutch drag 22

Gear selection problems
Gear lever does not return 23
Gear selection difficult or impossible 24
Jumping out of gear 25

Overselection 26

Abnormal engine noise
Knocking or pinking 27
Piston slap or rattling from cylinder 28
Other noises 29

Abnormal transmission noise
Clutch noise 30
Transmission noise 31

Exhaust smokes excessively
White/blue smoke (caused by oil burning) 32
Black smoke (caused by over-rich mixture) 33

Poor handling or roadholding
Directional instability 34
Steering bias to left or right 35
Handlebar vibrates or oscillates 36
Poor front fork performance 37
Front fork judder when braking 38
Poor rear suspension performance 39

Abnormal frame and suspension noise
Front end noise 40
Rear suspension noise 41

Brake problems
Brakes are spongy or ineffective – disc brakes 42
Brakes drag – disc brakes 43
Brake lever or pedal pulsates in operation – disc brakes ... 44
Disc brake noise 45
Brakes are spongy or ineffective – drum brakes 46
Brake drag – drum brakes 47
Brake lever or pedal pulsates in operation – drum brakes . 48
Drum brake noise 49
Brake induced fork judder 50

Electrical problems
Battery dead or weak 51
Battery overcharged 52
Total electrical failure 53
Circuit failure 54
Bulbs blowing repeatedly 55

---

1 Introduction

This Section provides as easy reference-guide to the more common ailments that are likely to afflict your machine. Obviously, the opportunities are almost limitless for faults to occur as a result of obscure failures, and to try and cover all eventualities would require a book. Indeed, a number have been written on the subject.

Successful fault diagnosis is not a mysterious 'black art' but the application of a bit of knowledge combined with a systematic and logical approach to the problem. Approach any fault diagnosis by first accurately identifying the symptom and then checking through the list of possibile causes, starting with the simplest or most obvious and progressing in stages to the most complex. Take nothing for granted, but above all apply liberal quantities of common sense.

The main symptom of a fault is given in the text as a major heading below which are listed, as Sections headings, the various systems or areas which may contain the fault. Details of each possible cause for a fault and the remedial action to be taken are given, in brief, in the paragraphs below each Section heading. Further information should be sought in the relevant Chapter.

Engine does not start when turned over

2 No fuel flow to carburettor

● Fuel tank empty or level too low. Check that the tap is turned to 'On' or 'Reserve' position as required. If in doubt, prise off the fuel feed pipe at the carburettor end and check that fuel runs from pipe when the tap is turned on.
● Tank filler cap vent obstructed. This can prevent fuel from flowing into the carburettor float bowl bcause air cannot enter the fuel tank to replace it. The problem is more likely to appear when the machine is being ridden. Check by listening close to the filler cap and releasing it. A hissing noise indicates that a blockage is present. Remove the cap and clear the vent hole with wire or by using an air line from the inside of the cap.
● Fuel tap or filter blocked. Blockage may be due to accumulation of rust or paint flakes from the tank's inner surface or of foreign matter from contaminated fuel. Remove the tap and clean it and the filter. Look also for water droplets in the fuel.
● Fuel line blocked. Blockage of the fuel line is more likely to result from a kink in the line rather than the accumulation of debris.

3 Fuel not reaching cylinder

● Float chamber not filling. Caused by float needle or floats sticking in up position. This may occur after the machine has been left standing for an extended length of time allowing the fuel to evaporate. When this occurs a gummy residue is often left which hardens to a varnish-like substance. This condition may be worsened by corrosion and crystalline deposits produced prior to the total evaporation of contaminated fuel. Sticking of the float needle may also be caused by wear. In any case removal of the float chamber will be necessary for inspection and cleaning.
● Blockage in starting circuit, slow running circuit or jets. Blockage of these items may be attributable to debris from the fuel tank by-passing the filter system or to gumming up as described in paragraph 1. Water droplets in the fuel will also block jets and passages. The carburettor should be dismantled for cleaning.
● Fuel level too low. The fuel level in the float chamber is controlled by float height. The fuel level may increase with wear or damage but will never reduce, thus a low fuel level is an inherent rather than developing condition. Check the float height and make any necessary adjustment.
● Oil blockage in fuel system or carburettor (petroil lubricated engines only). May arise when the machine has been parked for long periods and the residual petrol has evaporated. To rectify, dismantle and clean the carburettor and tap, flush the tank and fill with fresh petroil mixed in the correct proportions. This problem can be avoided by running the float bowl dry before the machine is stored for long periods. Do not attempt to use fuel which has become stale.

4 Engine flooding

● Float valve needle worn or stuck open. A piece of rust or other debris can prevent correct seating of the needle against the valve seat thereby permitting an uncontrolled flow of fuel. Similarly, a worn needle or needle seat will prevent valve closure. Dismantle the carburettor float bowl for cleaning and, if necessary, renewal of the worn components.
● Fuel level too high. The fuel level is controlled by the float height which may increase due to wear of the float needle, pivot pin or operating tang. Check the float height, and make any necessary adjustment. A leaking float will cause an increase in fuel level, and thus should be renewed.
● Cold starting mechanism. Check the choke (starter mechanism) for correct operation. If the mechanism jams in the 'On' position subsequent starting of a hot engine will be difficult.
● Blocked air filter. A badly restricted air filter will cause flooding. Check the filter and clean or renew as required. A collapsed inlet hose will have a similar effect. Check that the air filter inlet has not become blocked by a rag or similar item.

5 No spark at plug

● Ignition switch not on.
● Engine stop switch off.
● Fuse blown. Check fuse for ignition circuit. See wiring diagram.
● Spark plug dirty, oiled or 'whiskered'. Because the induction mixture of a two-stroke engine is inclined to be of a rather oily nature it is comparatively easy to foul the plug electrodes, especially where there have been repeated attempts to start the engine. A machine used for short journeys will be more prone to fouling because the engine may never reach full operating temperature, and the deposits will not burn off. On rare occasions a change of plug grade may be required but the advice of a dealer should be sought before making such a change. 'Whiskering' is a comparatively rare occurrence on modern machines but may be encountered where pre-mixed petrol and oil (petroil) lubrication is employed. An electrode deposit in the form of a barely visible filament across the plug electrodes can short circuit the plug and prevent its sparking. On all two-stroke machines it is a sound precaution to carry a new spare spark plug for substitution in the event of fouling problems.
● Spark plug failure. Clean the spark plug thoroughly and reset the electrode gap. Refer to the spark plug section and the colour condition guide in Chapter 4. If the spark plug shorts internally or has sustained visible damage to the electrodes, core or ceramic insulator it should be renewed. On rare occasions a plug that appears to spark vigorously will fail to do so when refitted to the engine and subjected to the compression pressure in the cylinder.
● Spark plug cap or high tension (HT) lead faulty. Check condition and security. Replace if deterioration is evident. Most spark plugs have an internal resistor designed to inhibit electrical interference with radio and television sets. On rare occasions the resistor may break down, thus preventing sparking. If this is suspected, fit a new cap as a precaution.
● Spark plug cap loose. Check that the spark plug cap fits securely over the plug and, where fitted, the screwed terminal on the plug end is secure.
● Shorting due to moisture. Certain parts of the ignition system are susceptible to shorting when the machine is ridden or parked in wet weather. Check particularly the area from the spark plug cap back to the ignition coil. A water dispersant spray may be used to dry out waterlogged components. Recurrence of the problem can be prevented by using an ignition sealant spray after drying out and cleaning.
● Ignition or stop switch shorted. May be caused by water corrosion or wear. Water dispersant and contact cleaning sprays may be used. If this fails to overcome the problem dismantling and visual inspection of the switches will be required.
● Shorting or open circuit in wiring. Failure in any wire connecting any of the ignition components will cause ignition malfunction. Check also that all connections are clean, dry and tight.
● Ignition coil failure. Check the coil, referring to Chapter 4.
● Pulser coil, source coil or CDI unit defective. See Chapter 4.

6 Weak spark at plug

● Feeble sparking at the plug may be caused by any of the faults mentioned in the preceding Section other than those items in the first three paragraphs. Check first the spark plug, this being the most likely culprit.

7 Compression low

● Spark plug loose. This will be self-evident on inspection, and may be accompanied by a hissing noise when the engine is turned over. Remove the plug and check that the threads in the cylinder head are not damaged. Check also that the plug sealing washer is in good condition.
● Cylinder head gasket leaking. This condition is often accompanied by a high pitched squeak from around the cylinder head and oil loss, and may be caused by insufficiently tightened cylinder head fasteners, a warped cylinder head or mechanical failure of the gasket material. Re-torquing the fasteners to the correct specification may seal the leak in some instances but if damage has occurred this course of action will provide, at best, only a temporary cure.
● Low crankcase compression. This can be caused by worn main bearings and seals and will upset the incoming fuel/air mixture. A good seal in these areas is essential on any two-stroke engine.
● Worn disc valve. Disc valve wear is not common, but will cause similar symptoms to those described above. Overhaul will be necessary.
● Piston rings sticking or broken. Sticking of the piston rings may be caused by seizure due to lack of lubrication or heating as a result of poor carburation or incorrect fuel type. Gumming of the rings may result from lack of use, or carbon deposits in the ring grooves. Broken rings result from over-revving, over-heating or general wear. In either case a top-end overhaul will be required.
● Worn or damaged decompressor valve. Some mopeds are fitted with a small cable-operated poppet valve in the cylinder head for use when starting or as an engine stop device. If this fails to seat properly or becomes jammed open compression will be lost and in extreme case starting will be impossible. Access to the valve requires the removal of the cylinder head.

Engine stalls after starting

8 General causes

● Improper cold start mechanism operation. Check that the operating controls function smoothly and, where applicable, are correctly adjusted. A cold engine may not require application of an enriched mixture to start initially but may baulk without choke once firing. Likewise a hot engine may start with an enriched mixture but will stop almost immediately if the choke is inadvertently in operation.
● Ignition malfunction. See Section 9. Weak spark at plug.
● Carburettor incorrectly adjusted. Maladjustment of the mixture strength or idle speed may cause the engine to stop immediately after starting. See Chapter 2.
● Fuel contamination. Check for filter blockage by debris or water which reduces, but does not completely stop, fuel flow, or blockage of the slow speed circuit in the carburettor by the same agents. If water is present it can often be seen as droplets in the bottom of the float bowl. Clean the filter and, where water is in evidence, drain and flush the fuel tank and float bowl.
● Intake air leak. Check for security of the carburettor mounting and hose connections, and for cracks or splits in the hoses. Check also that the carburettor top is secure and that the vacuum gauge adaptor plug (where fitted) is tight.
● Air filter blocked or omitted. A blocked filter will cause an over-rich mixture; the omission of a filter will cause an excessively weak mixture. Both conditions will have a detrimental affect on carburation. Clean or renew the filter as necessary.
● Fuel filler cap air vent blocked. Usually caused by dirt or water. Clean the vent orifice.
● Choked exhaust system. Caused by excessive carbon build-

up in the system, particularly around the silencer baffles. In many cases these can be detached for cleaning, though mopeds have one-piece systems which require a rather different approach. Refer to Chapter 2 for further information.
● Excessive carbon build-up in the engine. This can result from failure to decarbonise the engine at the specified interval or through excessive oil consumption. On pump-fed engines check pump adjustment. On pre-mix (petroil) systems check that oil is mixed in the recommended ratio.

Poor running at idle and low speed

9 Weak spark at plug or erratic firing

● Battery voltage low. In certain conditions low battery charge, especially when coupled with a badly sulphated battery, may result in misfiring. If the battery is in good general condition it should be recharged; an old battery suffering from sulphated plates should be renewed.
● Spark plug fouled, faulty or incorrectly adjusted. See Section 4 or refer to Chapter 4.
● Spark plug cap or high tension lead shorting. Check the condition of both these items ensuring that they are in good condition and dry and that the cap is fitted correctly.
● Spark plug type incorrect. Fit plug of correct type and heat range as given in Specifications. In certain conditions a plug of hotter or colder type may be required for normal running.
● Pulser coil, source coil or CDI unit faulty. See Chapter 4.
● Ignition timing incorrect. Check the ignition timing statically and dynamically, ensuring that the advance is functioning correctly.
● Faulty ignition coil. Partial failure of the coil internal insulation will diminish the performance of the coil. No repair is possible, a new component must be fitted.

10 Fuel/air mixture incorrect

● Intake air leak. Check carburettor mountings and air cleaner hoses for security and signs of splitting. Ensure that carburettor top is tight and that the vacuum gauge take-off plug (where fitted) is tight.
● Mixture strength incorrect. Adjust slow running mixture strength using pilot adjustment screw.
● Carburettor synchronisation.
● Pilot jet or slow running circuit blocked. The carburettor should be removed and dismantled for thorough cleaning. Blow through all jets and air passages with compressed air to clear obstructions.
● Air cleaner clogged or omitted. Clean or fit air cleaner element as necessary. Check also that the element and air filter cover are correctly seated.
● Cold start mechanism in operation. Check that the choke has not been left on inadvertently and the operation is correct. Where applicable check the operating cable free play.
● Fuel level too high or too low. Check the float height and adjust as necessary. See Section 3 or 4.
● Fuel tank air vent obstructed. Obstructions usually caused by dirt or water. Clean vent orifice.

11 Compression low

● See Section 7.

Acceleration poor

12 General causes

● All items as for previous Section.
● Choked air filter. Failure to keep the air filter element clean

will allow the build-up of dirt with proportional loss of performance. In extreme cases of neglect acceleration will suffer.

● Choked exhaust system. This can result from failure to remove accumulations of carbon from the silencer baffles at the prescribed intervals. The increased back pressure will make the machine noticeably sluggish. Refer to Chapter 2 for further information on decarbonisation.

● Excessive carbon build-up in the engine. This can result from failure to decarbonise the engine at the specified interval or through excessive oil consumption. On pump-fed engines check pump adjustment. On pre-mix (petroil) systems check that oil is mixed in the recommended ratio.

● Ignition timing incorrect. Check the ignition timing as described in Chapter 4. Where no provision for adjustment exists, test the electronic ignition components and renew as required.

● Carburation fault. See Section 10.

● Mechanical resistance. Check that the brakes are not binding. On small machines in particular note that the increased rolling resistance caused by under-inflated tyres may impede acceleration.

Poor running or lack of power at high speeds

13 Weak spark at plug or erratic firing

● All items as for Section 9.

● HT lead insulation failure. Insulation failure of the HT lead and spark plug cap due to old age or damage can cause shorting when the engine is driven hard. This condition may be less noticeable, or not noticeable at all at lower engine speeds.

14 Fuel/air mixture incorrect

● All items as for Section 10, with the exception of items relative exclusively to low speed running.

● Main jet blocked. Debris from contaminated fuel, or from the fuel tank, and water in the fuel can block the main jet. Clean the fuel filter, the float bowl area, and if water is present, flush and refill the fuel tank.

● Main jet is the wrong size. The standard carburettor jetting is for sea level atmospheric pressure. For high altitudes, usually above 5000 ft, a smaller main jet will be required.

● Jet needle and needle jet worn. These can be renewed individually but should be renewed as a pair. Renewal of both items requires partial dismantling of the carburettor.

● Air bleed holes blocked. Dismantle carburettor and use compressed air to blow out all air passages.

● Reduced fuel flow. A reduction in the maximum fuel flow from the fuel tank to the carburettor will cause fuel starvation, proportionate to the engine speed. Check for blockages through debris or a kinked fuel line.

● Vacuum diaphragm split. Renew.

15 Compression low

● See Section 7.

Knocking or pinking

16 General causes

● Carbon build-up in combustion chamber. After high mileages have been covered large accumulation of carbon may occur. This may glow red hot and cause premature ignition of the fuel/air mixture, in advance of normal firing by the spark plug. Cylinder head removal will be required to allow inspection and cleaning.

● Fuel incorrect. A low grade fuel, or one of poor quality may

result in compression induced detonation of the fuel resulting in knocking and pinking noises. Old fuel can cause similar problems. A too highly leaded fuel will reduce detonation but will accelerate deposit formation in the combustion chamber and may lead to early pre-ignition as described in item 1.

● Spark plug heat range incorrect. Uncontrolled pre-ignition can result from the use of a spark plug the heat range of which is too hot.

● Weak mixture. Overheating of the engine due to a weak mixture can result in pre-ignition occurring where it would not occur when engine temperature was within normal limits. Maladjustment, blocked jets or passages and air leaks can cause this condition.

Overheating

17 Firing incorrect

● Spark plug fouled, defective or maladjusted. See Section 5.

● Spark plug type incorrect. Refer to the Specifications and ensure that the correct plug type is fitted.

● Incorrect ignition timing. Timing that is far too much advanced or far too much retarded will cause overheating. Check the ignition timing is correct.

18 Fuel/air mixture incorrect

● Slow speed mixture strength incorrect. Adjust pilot air screw.

● Main jet wrong size. The carburettor is jetted for sea level atmospheric conditions. For high altitudes, usually above 5000 ft, a smaller main jet will be required.

● Air filter badly fitted or omitted. Check that the filter element is in place and that it and the air filter box cover are sealing correctly. Any leaks will cause a weak mixture.

● Induction air leaks. Check the security of the carburettor mountings and hose connections, and for cracks and splits in the hoses. Check also that the carburettor top is secure and that the vacuum gauge adaptor plug (where fitted) is tight.

● Fuel level too low. See Section 3.

● Fuel tank filler cap air vent obstructed. Clear blockage.

19 Lubrication inadequate

● Petrol/oil mixture incorrect. The proportion of oil mixed with the petrol in the tank is critical if the engine is to perform correctly. Too little oil will leave the reciprocating parts and bearings poorly lubricated and overheating will occur. In extreme case the engine will seize. Conversely, too much oil will effectively displace a similar amount of petrol. Though this does not often cause overheating in practice it is possible that the resultant weak mixture may cause overheating. It will inevitably cause a loss of power and excessive exhaust smoke.

● Oil pump settings incorrect. The oil pump settings are of great importance since the quantities of oil being injected are very small. Any variation in oil delivery will have a significant effect on the engine. Refer to Chapter 3 for further information.

● Oil tank empty or low. This will have disastrous consequences if left unnoticed. Check and replenish tank regularly.

● Transmission oil low or worn out. Check the level regularly and investigate any loss of oil. If the oil level drops with no sign of external leakage it is likely that the crankshaft main bearing oil seals are worn, allowing transmission oil to be drawn into the crankcase during induction.

20 Miscellaneous causes

● Engine fins clogged. A build-up of mud in the cylinder head and cylinder barrel cooling fins will decrease the cooling capabilities of the fins. Clean the fins as required.

● Radiator fins clogged. Accumulated debris in the radiator core will gradually reduce its ability to dissipate heat generated by the engine. It is worth noting that during the summer months dead insects can cause as many problems in this respect as road dirt and mud during the winter. Cleaning is best carried out by dislodging the debris with a high pressure hose from the back of the radiator. Once cleaned it is worth painting the matrix with a heat-dispersant matt black paint both to assist cooling and to prevent external corrosion. The fitting of some sort of mesh guard will help prevent the fins from becoming clogged, but make sure that this does not itself prevent adequate cooling.

Clutch operating problems

21 Clutch slip

● No clutch lever play. Adjust clutch lever end play according to the procedure in Chapter 1.
● Friction plates worn or warped. Overhaul clutch assembly, replacing plates out of specification.
● Steel plates worn or warped. Overhaul clutch assembly, replacing plates out of specification.
● Clutch spring broken or worn. Old or heat-damaged (from slipping clutch) springs should be replaced with new ones.
● Clutch release not adjusted properly. See the adjustments section of Chapter 1.
● Clutch inner cable snagging. Caused by a frayed cable or kinked outer cable. Replace the cable with a new one. Repair of a frayed cable is not advised.
● Clutch release mechanism defective. Worn or damaged parts in the clutch release mechanism could include the shaft, cam, actuating arm or pivot. Replace parts as necessary.
● Clutch hub and outer drum worn. Severe indentation by the clutch plate tangs of the channels in the hub and drum will cause snagging of the plates preventing correct engagement. If this damage occurs, renewal of the worn components is required.
● Lubricant incorrect. Use of a transmission lubricant other than that specified may allow the plates to slip.

22 Clutch drag

● Clutch lever excessive. Adjust lever at bars or at cable end if necessary.
● Clutch plates warped or damaged. This will cause a drag on the clutch, causing the machine to creep. Overhaul clutch assembly.
● Clutch spring tension uneven. Usually caused by a sagged or broken spring. Check and replace springs.
● Transmission oil deteriorated. Badly contaminated transmission oil and a heavy deposit of oil sludge on the plates will cause plate sticking. The oil recommended for this machine is of the detergent type, therefore it is unlikely that this problem will arise unless regular oil changes are neglected.
● Transmission oil viscosity too high. Drag in the plates will result from the use of an oil with too high a viscosity. In very cold weather clutch drag may occur until the engine has reached operating temperature.
● Clutch hub and outer drum worn. Indentation by the clutch plate tangs of the channels in the hub and drum will prevent easy plate disengagement. If the damage is light the affected areas may be dressed with a fine file. More pronounced damage will necessitate renewal of the components.
● Clutch housing seized to shaft. Lack of lubrication, severe wear or damage can cause the housing to seize to the shaft. Overhaul of the clutch, and perhaps the transmission, may be necessary to repair damage.
● Clutch release mechanism defective. Worn or damaged release mechanism parts can stick and fail to provide leverage.

Overhaul clutch cover components.
● Loose clutch hub nut. Causes drum and hub misalignment, putting a drag on the engine. Engagement adjustment continually varies. Overhaul clutch assembly.

Gear selection problems

23 Gear lever does not return

● Weak or broken centraliser spring. Renew the spring.
● Gearchange shaft bent or seized. Distortion of the gearchange shaft often occurs if the machine is dropped heavily on the gear lever. Provided that damage is not severe straightening of the shaft is permissible.

24 Gear selection difficult or impossible

● Clutch not disengaging fully. See Section 22.
● Gearchange shaft bent. This often occurs if the machine is dropped heavily on the gear lever. Straightening of the shaft is permissible if the damage is not too great.
● Gearchange arms, pawls or pins worn or damaged. Wear or breakage of any of these items may cause difficulty in selecting one or more gears. Overhaul the selector mechanism.
● Gearchange shaft centraliser spring maladjusted. This is often characterised by difficulties in changing up or down, but rarely in both directions. Adjust the centraliser anchor bolt as described in Chapter 1.
● Gearchange drum stopper cam or detent plunger damage. Failure, rather than wear of these items may jam the drum thereby preventing gearchanging or causing false selection at high speed.
● Selector forks bent or seized. This can be caused by dropping the machine heavily on the gearchange lever or as a result of lack of lubrication. Though rare, bending of a shaft can result from a missed gearchange or false selection at high speed.
● Selector fork end and pin wear. Pronounced wear of these items and the grooves in the gearchange drum can lead to imprecise selection and, eventually, no selection. Renewal of the worn components will be required.
● Structural failure. Failure of any one component of the selector rod and change mechanism will result in improper or fouled gear selection.

25 Jumping out of gear

● Detent plunger assembly worn or damaged. Wear of the plunger and the cam with which it locates and breakage of the detent spring can cause imprecise gear selection resulting in jumping out of gear. Renew the damaged components.
● Gear pinion dogs worn or damaged. Rounding off the dog edges and the mating recesses in adjacent pinion can lead to jumping out of gear when under load. The gears should be inspected and renewed. Attempting to reprofile the dogs is not recommended.
● Selector forks, gearchange drum and pinion grooves worn. Extreme wear of these interconnected items can occur after high mileages especially when lubrication has been neglected. The worn components must be renewed.
● Gear pinions, bushes and shafts worn. Renew the worn components.
● Bent gearchange shaft. Often caused by dropping the machine on the gear lever.
● Gear pinion tooth broken. Chipped teeth are unlikely to cause jumping out of gear once the gear has been selected fully; a tooth which is completely broken off, however, may cause problems in this respect and in any event will cause transmission noise.

26 Overselection

● Pawl spring weak or broken. Renew the spring.
● Detent plunger worn or broken. Renew the damaged items.
● Stopper arm spring worn or broken. Renew the spring.
● Gearchange arm stop pads worn. Repairs can be made by welding and reprofiling with a file.
● Selector limiter claw components (where fitted) worn or damaged. Renew the damaged items.

Abnormal engine noise

27 Knocking or pinking

● See Section 16.

28 Piston slap or rattling from cylinder

● Cylinder bore/piston clearance excessive. Resulting from wear, partial seizure or improper boring during overhaul. This condition can often be heard as a high, rapid tapping noise when the engine is under little or no load, particularly when power is just beginning to be applied. Reboring to the next correct oversize should be carried out and a new oversize piston fitted.
● Connecting rod bent. This can be caused by over-revving, trying to start a very badly flooded engine (resulting in a hydraulic lock in the cylinder) or by earlier mechanical failure. Attempts at straightening a bent connecting rod from a high performance engine are not recommended. Careful inspection of the crankshaft should be made before renewing the damaged connecting rod.
● Gudgeon pin, piston boss bore or small-end bearing wear or seizure. Excess clearance or partial seizure between normal moving parts of these items can cause continuous or intermittent tapping noises. Rapid wear or seizure is caused by lubrication starvation.
● Piston rings worn, broken or sticking. Renew the rings after careful inspection of the piston and bore.

29 Other noises

● Big-end bearing wear. A pronounced knock from within the crankcase which worsens rapidly is indicative of big-end bearing failure as a result of extreme normal wear or lubrication failure. Remedial action in the form of a bottom end overhaul should be taken; continuing to run the engine will lead to further damage including the possibility of connecting rod breakage.
● Main bearing failure. Extreme normal wear or failure of the main bearings is characteristically accompanied by a rumble from the crankcase and vibration felt through the frame and footrests. Renew the worn bearings and carry out a very careful examination of the crankshaft.
● Crankshaft excessively out of true. A bent crank may result from over-revving or damage from an upper cylinder component or gearbox failure. Damage can also result from dropping the machine on either crankshaft end. Straightening of the crankshaft is not possible in normal circumstances; a replacement item should be fitted.
● Engine mounting loose. Tighten all the engine mounting nuts and bolts.
● Cylinder head gasket leaking. The noise most often associated with a leaking head gasket is a high pitched squeaking, although any other noise consistent with gas being forced out under pressure from a small orifice can also be emitted. Gasket leakage is often accompanied by oil seepage from around the mating joint or from the cylinder head holding down bolts and nuts. Leakage results from insufficient or uneven tightening of the cylinder head fasteners, or from random mechanical failure. Retightening to the correct torque figure will, at best, only provide a temporary cure. The gasket should be renewed at the earliest opportunity.
● Exhaust system leakage. Popping or crackling in the exhaust system, particularly when it occurs with the engine on the overrun, indicates a poor joint either at the cylinder port or at the exhaust pipe/silencer connection. Failure of the gasket or looseness of the clamp should be looked for.

Abnormal transmission noise

30 Clutch noise

● Clutch outer drum/friction plate tang clearance excessive.
● Clutch outer drum/spacer clearance excessive.
● Clutch outer drum/thrust washer clearance excessive.
● Primary drive gear teeth worn or damaged.
● Clutch shock absorber assembly worn or damaged.

31 Transmission noise

● Bearing or bushes worn or damaged. Renew the affected components.
● Gear pinions worn or chipped. Renew the gear pinions.
● Metal chips jammed in gear teeth. This can occur when pieces of metal from any failed component are picked up by a meshing pinion. The condition will lead to rapid bearing wear or early gear failure.
● Engine/transmission oil level too low. Top up immediately to prevent damage to gearbox and engine.
● Gearchange mechanism worn or damaged. Wear or failure of certain items in the selection and change components can induce mis-selection of gears (see Section 24) where incipient engagement of more than one gear set is promoted. Remedial action, by the overhaul of the gearbox, should be taken without delay.
● Loose gearbox chain sprocket. Remove the sprocket and check for impact damage to the splines of the sprocket and shaft. Excessive slack between the splines will promote loosening of the securing nut; renewal of the worn components is required. When retightening the nut ensure that it is tightened fully and that, where fitted, the lock washer is bent up against one flat of the nut.
● Chain snagging on cases or cycle parts. A badly worn chain or one that is excessively loose may snag or smack against adjacent components.

Exhaust smokes excessively

32 White/blue smoke (caused by oil burning)

● Piston rings worn or broken. Breakage or wear of any ring, but particularly the oil control ring, will allow engine oil past the piston into the combustion chamber. Overhaul the cylinder barrel and piston.
● Cylinder cracked, worn or scored. These conditions may be caused by overheating, lack of lubrication, component failure or advanced normal wear. The cylinder barrel should be renewed or rebored and the next oversize piston fitted.
● Petrol/oil ratio incorrect. Ensure that oil is mixed with the petrol in the correct ratio. The manufacturer's recommendation must be adhered to if excessive smoking or under-lubrication is to be avoided.
● Oil pump settings incorrect. Check and reset the oil pump as described in Chapter 2.
● Crankshaft main bearing oil seals worn. Wear in the main bearing oil seals, often in conjunction with wear in the bearings themselves, can allow transmission oil to find its way into the

crankcase and thence to the combustion chamber. This condition is often indicated by a mysterious drop in the transmission oil level with no sign of external leakage.

● Accumulated oil deposits in exhaust system. If the machine is used for short journeys only it is possible for the oil residue in the exhaust gases to condense in the relatively cool silencer. If the machine is then taken for a longer run in hot weather, the accumulated oil will burn off producing ominous smoke from the exhaust.

33 Black smoke (caused by over-rich mixture)

● Air filter element clogged. Clean or renew the element.
● Main jet loose or too large. Remove the float chamber to check for tightness of the jet. If the machine is used at high altitudes rejetting will be required to compensate for the lower atmospheric pressure.
● Cold start mechanism jammed on. Check that the mechanism works smoothly and correctly and that, where fitted, the operating cable is lubricated and not snagged.
● Fuel level too high. The fuel level is controlled by the float height which can increase as a result of wear or damage. Remove the float bowl and check the float height. Check also that floats have not punctured; a punctured float will lose buoyancy and allow an increased fuel level.
● Float valve needle stuck open. Caused by dirt or a worn valve. Clean the float chamber or renew the needle and, if necessary, the valve seat.

Poor handling or roadholding

34 Directional instability

● Steering head bearing adjustment too tight. This will cause rolling or weaving at low speeds. Re-adjust the bearings.
● Steering head bearing worn or damaged. Correct adjustment of the bearing will prove impossible to achieve if wear or damage has occurred. Inconsistent handling will occur including rolling or weaving at low speed and poor directional control at indeterminate higher speeds. The steering head bearing should be dismantled for inspection and renewed if required. Lubrication should also be carried out.
● Bearing races pitted or dented. Impact damage caused, perhaps, by an accident or riding over a pot-hole can cause indentation of the bearing, usually in one position. This should be noted as notchiness when the handlebars are turned. Renew and lubricate the bearings.
● Steering stem bent. This will occur only if the machine is subjected to a high impact such as hitting a curb or a pot-hole. The lower yoke/stem should be renewed; do not attempt to straighten the stem.
● Front or rear tyre pressures too low.
● Front or rear tyre worn. General instability, high speed wobbles and skipping over white lines indicates that tyre renewal may be required. Tyre induced problems, in some machine/tyre combinations, can occur even when the tyre in question is by no means fully worn.
● Swinging arm bearings worn. Difficulties in holding line, particularly when cornering or when changing power settings indicates wear in the swinging arm bearings. The swinging arm should be removed from the machine and the bearings renewed.
● Swinging arm flexing. The symptoms given in the preceding paragraph will also occur if the swinging arm fork flexes badly. This can be caused by structural weakness as a result of corrosion, fatigue or impact damage, or because the rear wheel spindle is slack.
● Wheel bearings worn. Renew the worn bearings.
● Loose wheel spokes. The spokes should be tightened evenly to maintain tension and trueness of the rim.

● Tyres unsuitable for machine. Not all available tyres will suit the characteristics of the frame and suspension, indeed, some tyres or tyre combinations may cause a transformation in the handling characteristics. If handling problems occur immediately after changing to a new tyre type or make, revert to the original tyres to see whether an improvement can be noted. In some instances a change to what are, in fact, suitable tyres may give rise to handling deficiences. In this case a thorough check should be made of all frame and suspension items which affect stability.

35 Steering bias to left or right

● Rear wheel out of alignment. Caused by uneven adjustment of chain tensioner adjusters allowing the wheel to be askew in the fork ends. A bent rear wheel spindle will also misalign the wheel in the swinging arm.
● Wheels out of alignment. This can be caused by impact damage to the frame, swinging arm, wheel spindles or front forks. Although occasionally a result of material failure or corrosion it is usually as a result of a crash.
● Front forks twisted in the steering yokes. A light impact, for instance with a pot-hole or low curb, can twist the fork legs in the steering yokes without causing structural damage to the fork legs or the yokes themselves. Re-alignment can be made by loosening the yoke pinch bolts, wheel spindle and mudguard bolts. Re-align the wheel with the handlebars and tighten the bolts working upwards from the wheel spindle. This action should be carried out only when there is no chance that structural damage has occurred.

36 Handlebar vibrates or oscillates

● Tyres worn or out of balance. Either condition, particularly in the front tyre, will promote shaking of the fork assembly and thus the handlebars. A sudden onset of shaking can result if a balance weight is displaced during use.
● Tyres badly positioned on the wheel rims. A moulded line on each wall of a tyre is provided to allow visual verification that the tyre is correctly positioned on the rim. A check can be made by rotating the tyre; any misalignment will be immediately obvious.
● Wheels rims warped or damaged. Inspect the wheels for runout as described in Chapter 6.
● Swinging arm bearings worn. Renew the bearings.
● Wheel bearings worn. Renew the bearings.
● Steering head bearings incorrectly adjusted. Vibration is more likely to result from bearings which are too loose rather than too tight. Re-adjust the bearings.
● Loosen fork component fasteners. Loose nuts and bolts holding the fork legs, wheel spindle, mudguards or steering stem can promote shaking at the handlebars. Fasteners on running gear such as the forks and suspension should be check tightened occasionally to prevent dangerous looseness of components occurring.
● Engine mounting bolts loose. Tighten all fasteners.

37 Poor front fork performance

● Damping fluid level incorrect. If the fluid level is too low poor suspension control will occur resulting in a general impairment of roadholding and early loss of tyre adhesion when cornering and braking. Too much oil is unlikely to change the fork characteristics unless severe overfilling occurs when the fork action will become stiffer and oil seal failure may occur.
● Damping oil viscosity incorrect. The damping action of the fork is directly related to the viscosity of the damping oil. The lighter the oil used, the less will be the damping action

imparted. For general use, use the recommended viscosity of oil, changing to a slightly higher or heavier oil only when a change in damping characteristic is required. Overworked oil, or oil contaminated with water which has found its way past the seals, should be renewed to restore the correct damping performance and to prevent bottoming of the forks.

● Damping components worn or corroded. Advanced normal wear of the fork internals is unlikely to occur until a very high mileage has been covered. Continual use of the machine with damaged oil seals which allows the ingress of water, or neglect, will lead to rapid corrosion and wear. Dismantle the forks for inspection and overhaul.

● Weak fork springs. Progressive fatigue of the fork springs, resulting in a reduced spring free length, will occur after extensive use. This condition will promote excessive fork dive under braking, and in its advanced form will reduce the at-rest extended length of the forks and thus the fork geometry. Renewal of the springs as a pair is the only satisfactory course of action.

● Bent stanchions or corroded stanchions. Both conditions will prevent correct telescoping of the fork legs, and in an advanced state can cause sticking of the fork in one position. In a mild form corrosion will cause stiction of the fork thereby increasing the time the suspension takes to react to an uneven road surface. Bent fork stanchions should be attended to immediately because they indicate that impact damage has occurred, and there is a danger that the forks will fail with disastrous consequences.

38 Front fork judder when braking (see also Section 41)

● Wear between the fork stanchions and the fork legs. Renewal of the affected components is required.
● Slack steering head bearings. Re-adjust the bearings.
● Warped brake disc or drum. If irregular braking action occurs fork judder can be induced in what are normally serviceable forks. Renew the damaged brake components.

39 Poor rear suspension performances

● Rear suspension unit damper worn out or leaking. The damping performance of most rear suspension units falls off with age. This is a gradual process, and thus may not be immediately obvious. Indications of poor damping include hopping of the rear end when cornering or braking, and a general loss of positive stability.
● Weak rear springs. If the suspension unit springs fatigue they will promote excessive pitching of the machine and reduce the ground clearance when cornering. Although replacement springs are available separately from the rear suspension damper unit it is probable that if spring fatigue has occurred the damper units will also require renewal.
● Swinging arm flexing or bearings worn. See Sections 34 and 36.
● Bent suspension unit damper rod. This is likely to occur only if the machine is dropped or if seizure of the piston occurs. If either happens the suspension units should be renewed as a pair.

Abnormal frame and suspension noise

40 Front end noise

● Oil level low or too thin. This can cause a 'spurting' sound and is usually accompanied by irregular fork action.
● Spring weak or broken. Makes a clicking or scraping sound. Fork oil will have a lot of metal particles in it.
● Steering head bearings loose or damaged. Clicks when braking. Check, adjust or replace.

● Fork clamps loose. Make sure all fork clamp pinch bolts are tight.
● Fork stanchion bent. Good possibility if machine has been dropped. Repair or replace tube.

41 Rear suspension noise

● Fluid level too low. Leakage of a suspension unit, usually evident by oil on the outer surfaces, can cause a spurting noise. The suspension units should be renewed as a pair.
● Defective rear suspension unit with internal damage. Renew the suspension units as a pair.

Brake problems

42 Brakes are spongy or ineffective – disc brakes

● Air in brake circuit. This is only likely to happen in service due to neglect in checking the fluid level or because a leak has developed. The problem should be identified and the brake system bled of air.
● Pad worn. Check the pad wear against the wear lines provided and renew the pads if necessary.
● Contaminated pads. Cleaning pads which have been contaminated with oil, grease or brake fluid is unlikely to prove successful; the pads should be renewed.
● Pads glazed. This is usually caused by overheating. The surface of the pads may be roughened using glass-paper or a fine file.
● Brake fluid deterioration. A brake which on initial operation is firm but rapidly becomes spongy in use may be failing due to water contamination of the fluid. The fluid should be drained and then the system refilled and bled.
● Master cylinder seal failure. Wear or damage of master cylinder internal parts will prevent pressurisation of the brake fluid. Overhaul the master cylinder unit.
● Caliper seal failure. This will almost certainly be obvious by loss of fluid, a lowering of fluid in the master cylinder reservoir and contamination of the brake pads and caliper. Overhaul the caliper assembly.
● Brake lever or pedal improperly adjusted. Adjust the clearance between the lever end and master cylinder plunger to take up lost motion, as recommended in Routine maintenance.

43 Brakes drag – disc brakes

● Disc warped. The disc must be renewed.
● Caliper piston, caliper or pads corroded. The brake caliper assembly is vulnerable to corrosion due to water and dirt, and unless cleaned at regular intervals and lubricated in the recommended manner, will become sticky in operation.
● Piston seal deteriorated. The seal is designed to return the piston in the caliper to the retracted position when the brake is released. Wear or old age can affect this function. The caliper should be overhauled if this occurs.
● Brake pad damaged. Pad material separating from the backing plate due to wear or faulty manufacture. Renew the pads. Faulty installation of a pad also will cause dragging.
● Wheel spindle bent. The spindle may be straightened if no structural damage has occurred.
● Brake lever or pedal not returning. Check that the lever or pedal works smoothly throughout its operating range and does not snag on any adjacent cycle parts. Lubricate the pivot if necessary.
● Twisted caliper support bracket. This is likely to occur only after impact in an accident. No attempt should be made to re-align the caliper; the bracket should be renewed.

44 Brake lever or pedal pulsates in operation – disc brakes

● Disc warped or irregularly worn. The disc must be renewed.
● Wheel spindle bent. The spindle may be straightened provided no structural damage has occurred.

45 Disc brake noise

● Brake squeal. This can be caused by the omission or incorrect installation of the anti-squeal shim fitted to the rear of one pad. The arrow on the shim should face the direction of wheel normal rotation. Squealing can also be caused by dust on the pads, usually in combination with glazed pads, or other contamination from oil, grease, brake fluid or corrosion. Persistent squealing which cannot be traced to any of the normal causes can often be cured by applying a thin layer of high temperature silicone grease to the rear of the pads. Make absolutely certain that no grease is allowed to contaminate the braking surface of the pads.
● Glazed pads. This is usually caused by high temperatures or contamination. The pad surfaces may be roughened using glass-paper or a fine file. If this approach does not effect a cure the pads should be renewed.
● Disc warped. This can cause a chattering, clicking or intermittent squeal and is usually accompanied by a pulsating brake lever or pedal or uneven braking. The disc must be renewed.
● Brake pads fitted incorrectly or undersize. Longitudinal play in the pads due to omission of the locating springs (where fitted) or because pads of the wrong size have been fitted will cause a single tapping noise every time the brake is operated. Inspect the pads for correct installation and security.

46 Brakes are spongy or ineffective – drum brakes

● Brake cable deterioration. Damage to the outer cable by stretching or being trapped will give a spongy feel to the brake lever. The cable should be renewed. A cable which has become corroded due to old age or neglect of lubrication will partially seize making operation very heavy. Lubrication at this stage may overcome the problem but the fitting of a new cable is recommended.
● Worn brake linings. Determine lining wear using the external brake wear indicator on the brake backplate, or by removing the wheel and withdrawing the brake backplate. Renew the shoe/lining units as a pair if the linings are worn below the recommended limit.
● Worn brake camshaft. Wear between the camshaft and the bearing surface will reduce brake feel and reduce operating efficiency. Renewal of one or both items will be required to rectify the fault.
● Worn brake cam and shoe ends. Renew the worn components.
● Linings contaminated with dust or grease. Any accumulations of dust should be cleaned from the brake assembly and drum using a petrol dampened cloth. Do not blow or brush off the dust because it is asbestos based and thus harmful if inhaled. Light contamination from grease can be removed from the surface of the brake linings using a solvent; attempts at removing heavier contamination are less likely to be successful because some of the lubricant will have been absorbed by the lining material which will severely reduce the braking performance.

47 Brake drag – drum brakes

● Incorrect adjustment. Re-adjust the brake operating mechanism.

● Drum warped or oval. This can result from overheating or impact or uneven tension of the wheel spokes. The condition is difficult to correct, although if slight ovality only occurs, skimming the surface of the brake drum can provide a cure. This is work for a specialist engineer. Renewal of the complete wheel hub is normally the only satisfactory solution.
● Weak brake shoe return springs. This will prevent the brake lining/shoe units from pulling away from the drum surface once the brake is released. The springs should be renewed.
● Brake camshaft, lever pivot or cable poorly lubricated. Failure to attend to regular lubrication of these areas will increase operating resistance which, when compounded, may cause tardy operation and poor release movement.

48 Brake lever or pedal pulsates in operation – drum brakes

● Drums warped or oval. This can result from overheating or impact or uneven spoke tension. This condition is difficult to correct, although if slight ovality only occurs skimming the surface of the drum can provide a cure. This is work for a specialist engineer. Renewal of the hub is normally the only satisfactory solution.

49 Drum brake noise

● Drum warped or oval. This can cause intermittent rubbing of the brake linings against the drum. See the preceding Section.
● Brake linings glazed. This condition, usually accompanied by heavy lining dust contamination, often induces brake squeal. The surface of the linings may be roughened using glass-paper or a fine file.

50 Brake induced fork judder

● Worn front fork stanchions and legs, or worn or badly adjusted steering head bearings. These conditions, combined with uneven or pulsating braking as described in Sections 44 and 48 will induce more or less judder when the brakes are applied, dependent on the degree of wear and poor brake operation. Attention should be given to both areas of malfunction. See the relevant Sections.

Electrical problems

51 Battery dead or weak

● Battery faulty. Battery life should not be expected to exceed 3 to 4 years, particularly where a starter motor is used regularly. Gradual sulphation of the plates and sediment deposits will reduce the battery performance. Plate and insulator damage can often occur as a result of vibration. Complete power failure, or intermittent failure, may be due to a broken battery terminal. Lack of electrolyte will prevent the battery maintaining charge.
● Battery leads making poor contact. Remove the battery leads and clean them and the terminals, removing all traces of corrosion and tarnish. Reconnect the leads and apply a coating of petroleum jelly to the terminals.
● Load excessive. If additional items such as spot lamps, are fitted, which increase the total electrical load above the maximum alternator output, the battery will fail to maintain full charge. Reduce the electrical load to suit the electrical capacity.
● Regulator/rectifier failure.
● Alternator generating coils open-circuit or shorted.
● Charging circuit shorting or open circuit. This may be caused by frayed or broken wiring, dirty connectors or a faulty ignition switch. The system should be tested in a logical manner. See Section 54.

52 Battery overcharged

● Rectifier/regulator faulty. Overcharging is indicated if the battery becomes hot or it is noticed that the electrolyte level falls repeatedly between checks. In extreme cases the battery will boil causing corrosive gases and electrolyte to be emitted through the vent pipes.
● Battery wrongly matched to the electrical circuit. Ensure that the specified battery is fitted to the machine.

53 Total electrical failure

● Fuse blown. Check the main fuse. If a fault has occurred, it must be rectified before a new fuse is fitted.
● Battery faulty. See Section 51.
● Earth failure. Check that the frame main earth strap from the battery is securely affixed to the frame and is making a good contact.
● Ignition switch or power circuit failure. Check for current flow through the battery positive lead (red) to the ignition switch. Check the ignition switch for continuity.

54 Circuit failure

● Cable failure. Refer to the machine's wiring diagram and check the circuit for continuity. Open circuits are a result of loose or corroded connections, either at terminals or in-line connectors, or because of broken wires. Occasionally, the core of a wire will break without there being any apparent damage to the outer plastic cover.
● Switch failure. All switches may be checked for continuity in each switch position, after referring to the switch position boxes incorporated in the wiring diagram for the machine. Switch failure may be a result of mechanical breakage, corrosion or water.
● Fuse blown. Refer to the wiring diagram to check whether or not a circuit fuse is fitted. Replace the fuse, if blown, only after the fault has been identified and rectified.

55 Bulbs blowing repeatedly

● Vibration failure. This is often an inherent fault related to the natural vibration characteristics of the engine and frame and is, thus, difficult to resolve. Modifications of the lamp mounting, to change the damping characteristics, may help.
● Intermittent earth. Repeated failure of one bulb, particularly where the bulb is fed directly from the generator, indicates that a poor earth exists somewhere in the circuit. Check that a good contact is available at each earthing point in the circuit.
● Reduced voltage. Where a quartz-halogen bulb is fitted the voltage to the bulb should be maintained or early failure of the bulb will occur. Do not overload the system with additional electrical equipment in excess of the system's power capacity and ensure that all circuit connections are maintained clean and tight.

YAMAHA RD250 & 350 LC

Check list

Pre-ride checks

1 Check the fluid level and the operation of the front brake
2 Check the operation of the rear brake
3 Check the operation of the clutch
4 Check for full and free operation of the throttle twistgrip
5 Check the level of coolant
6 Check the level of engine oil
7 Check the level of transmission oil
8 Check that the final drive chain is correctly adjusted and lubricated
9 Inspect the wheels and tyres for damage and check the tyre pressures
10 Check that all items are securely fastened
11 Check the correct operation of the lights and instruments

Weekly or every 200 miles (300 km)

1 Top up the engine oil tank
2 Check the tyre pressures
3 Check the level of electrolyte in the battery
4 Check the level of coolant in the reservoir tank
5 Lubricate the exposed portions of the control cables
6 Lubricate and adjust the final drive chain
7 Safety check
8 Ensure that the lights, horn and indicators function properly

Monthly or every 1000 miles (1500 km)

1 Check the front brake pad wear and top up the hydraulic fluid
2 Adjust the rear brake
3 Lubricate the control cables and pivots
4 Check the level of transmission oil

Three monthly or every 2000 miles (3000 km)

1 Adjust the clutch operating mechanism
2 Adjust the carburettors
3 Check the oil pump setting
4 Remove and clean the air filter element
5 Checking the correct operation of the steering and suspension
6 Decarbonize the exhaust system
7 Remove and examine the spark plugs
8 Change the transmission oil
9 Remove, clean and lubricate the final drive chain
10 Check the machine for loose fittings or fasteners

Six monthly or every 4000 miles (6000 km)

1 Decarbonize the engine
2 Overhaul the cooling system
3 Examine the lighting system and electrical wiring
4 Check the ignition timing

Yearly or every 8000 miles (12 000 km)

1 Overhaul the carburettors
2 Examine the condition of the wheels and tyres
3 Change the front brake hydraulic fluid
4 Change the front fork oil
5 Grease all pivot points on the machine

Adjustment data

Tyre pressures	Front	Rear
Normal riding	25 psi (1.75 kg/cm²)	28 psi (2.00 kg/cm²)
High speed riding	28 psi (2.00 kg/cm²)	32 psi (2.25 kg/cm²)

Spark plug type	NGK B8ES
Spark plug gap	0.7 – 0.8 mm (0.028 – 0.032 in)
Idle speed	1200 ± 50 rpm

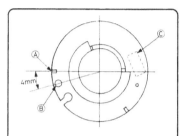

Oil pump pulley alignment marks

A
Standard alignment mark for 1980 models: RD250 LC up to engine number 4L1 100101 and RD350 LC up to engine number 4L0 100101

B
Modified alignment mark for the above models (see manual text)

C
Pulley identification mark for 1981 and 1982 models: RD250 LC from engine number 4L1 100101 and RD350 LC from engine number 4L0 100101 onwards. If pulley is marked 4L1 use mark A for 250 cc models, and mark B for 350 cc models.

Recommended lubricants

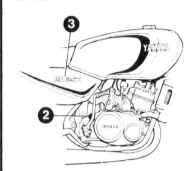

Component	Quantity	Type/viscosity
❶ Engine	1.6 lit (2.81 Imp pt)	Two-stroke oil
❷ Transmission	1.5 lit (2.64 Imp pt)	SAE 10W/30 engine oil
❸ Cooling system	200 cc (0.18 Imp qt)	50/50 mixture of distilled water and ethylene glycol antifreeze
❹ Front forks	140 ± 2.5 cc (4.93 ± 0.098 Imp fl oz)	SAE 10W/30 or a good quality fork oil
❺ Wheel bearings	As required	High melting point grease
❻ Steering head bearings	As required	High melting point grease
❼ Sub-frame pivot bolt	As required	High melting point grease
❽ Front brake	As required	DOT 3 or SAE J1703
❾ Pivot points	As required	High melting point grease
❿ Control cables	As required	Light machine oil
⓫ Final drive chain	As required	Aerosol chain lubricant

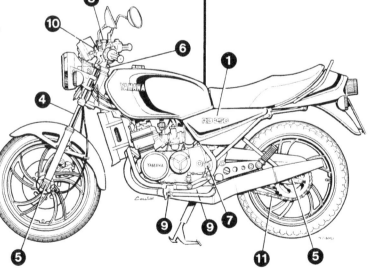

ROUTINE MAINTENANCE GUIDE

Routine maintenance

Introduction

Periodic routine maintenance is a continuous process that commences immediately the machine is used. It must be carried out at specified mileage recordings, or on a calendar basis if the machine is not used frequently, whichever is the sooner. Maintenance should be regarded as an insurance policy, to keep the machine in the peak of condition and to ensure long, trouble-free service. It has the additional benefit of giving early warning of any faults that may develop and will act as a regular safety check, to the obvious advantage of both rider and machine alike.

The various maintenance tasks are described under their respective mileage and calendar headings. Accompanying diagrams are provided, where necessary. It should be remembered that the interval between the various maintenance tasks serves only as a guide. As the machine gets older or is used under particularly adverse conditions, it would be advisable to reduce the period between each check.

For ease of reference each service operation is described in detail under the relevant heading. However, if further general information is required, it can be found within the manual under the pertinent section heading in the relevant Chapter.

In order that the routine maintenance tasks are carried out with as much ease as possible, it is essential that a good selection of general workshop tools are available.

Included in the kit must be a range of metric ring or combination spanners, a selection of crosshead screwdrivers and at least one pair of circlip pliers.

Additionally, owing to the extreme tightness of most casing screws on Japanese machines, an impact screwdriver, together with a choice of large or small crosshead screw bits, is absolutely indispensable. This is particularly so if the engine has not been dismantled since leaving the factory.

Pre-operation checks

Yamaha recommend that the checks shown in the accompanying table should be carried out each time the machine is used. Whilst this is sound advice it will obviously depend upon the use to which the machine is put. Few owners will feel inclined to carry out the complete sequence on every occasion. It is suggested that the list is tailored to individual usage but that all items are checked with reasonable frequency and before any long journeys.

a) Front brake. Check brake operation, free play and fluid level.
b) Rear brake. Check brake operation and free play.
c) Clutch. Check clutch operation and lever free play.
d) Throttle. Check for full and free operation of the twist grip.
e) Coolant. Check coolant level.
f) Engine oil. Check the oil level in the oil tank.
g) Transmission oil. Check the transmission oil level.
h) Final drive chain. Check the chain adjustment and lubrication.
i) Wheels and tyres. Check the wheels for damage and the tyres for damage or wear. Check the tyre pressures.
j) Fittings and fasteners. Check all items for security and damage.
k) Lights. Check that all illumination and warning lights are operating correctly.

Weekly or every 200 miles (300 km)

The following items should be checked on a weekly basis and as part of the pre-operation checklist shown above.

1 Topping up the oil tank

Engine lubrication is by pump injection fed by oil carried in a frame-mounted oil tank fitted behind the right-hand side panel. Although the machine is equipped with a low oil level warning lamp which will indicate the need for refilling it is better to check the oil level regularly and top up as required. Pull off the plastic side panel to expose the oil tank. The tank is constructed from a translucent white plastic which allows the oil level to be seen quite easily. To top up, release the single wing bolt and allow the tank to pivot outwards. The filler cap can now be pulled off and oil added. Use a good quality oil recommended for use in pump lubrication systems in air-cooled two-stroke engines.

2 Checking the tyre pressures

It is essential that the tyres are kept inflated to the correct pressure at all times. Under or over-inflated tyres can lead to accelerated rates of wear, and more importantly, can render the machine inherently unsafe. Whilst this may not be obvious during normal riding, it can become painfully and expensively so in an emergency situation, as the tyres' adhesion limits will be greatly reduced.

Check the tyre pressures with a pressure gauge that is known to be accurate. Always check the pressure when the tyres are cold. If the machine has travelled a number of miles, the tyres will have become hot and consequently the pressure will have increased. A false reading will therefore result.

It is recommended that a small pocket gauge is purchased and carried on the machine, as the readings on garage forecourt gauges can vary and may often be inaccurate.

The pressures given are those recommended for the tyres fitted as original equipment. If replacement tyres are purchased, the pressure settings may vary. Any reputable tyre distributor will be able to give this information.

Tyre pressures (cold)

	Front	Rear
Normal riding	25 psi (1.75 kg/cm^2)	28 psi (2.00 kg/cm^2)
High speed	28 psi (2.00 kg/cm^2)	32 psi (2.25 kg/cm^2)

3 Checking the electrolyte level

A Nippon Denso battery is fitted as standard equipment. It is of the normal lead-acid type and has a capacity of 5.5 Ah (Ampere-hours).

The transparent plastic case of the battery permits the upper and lower levels of the electrolyte to be observed when the left-hand side panel has been removed. Maintenance is normally limited to keeping the electrolyte level between the prescribed upper and lower limits and by making sure that the vent pipe is not blocked. The lead plates and their separators can be seen through the transparent case, a further guide to the general condition of the battery.

Unless acid is spilt, as may occur if the machine falls over, the electrolyte should always be topped up with distilled water, to restore the correct level. If acid is spilt on any part of the machine, it should be neutralised with an alkali such as washing soda and washed away with plenty of water, otherwise serious corrosion will occur. Top up with sulphuric acid of the correct specific gravity (1.260 – 1.280) only when spillage has occurred. Check that the vent pipe is well clear of the frame tubes or any of the other cycle parts, for obvious reasons.

4 Checking the coolant level

The cooling system is of the semi-sealed type and is unlikely to require frequent topping up. Thies does not mean, however, that regular checks should be neglected. The system employs a reservoir tank located next to the oil tank. This allows room for expansion when the engine becomes hot, the displaced water being drawn back into the radiator when it cools.

To check the level, remove the right-hand side panel. The reservoir tank has upper and lower level lines marked on its side, and the coolant level must be between these marks with the engine cold.

The mixture is a mixture of 50% soft or distilled water and 50% ethylene glycol antifreeze with corrosion inhibitors for use in aluminium engines. For topping up purposes, distilled water is best, or soft water where this is available. Those living in hard water areas should avoid the use of hard tap water because of the risk of scaling in the cooling system. **Clean** rainwater may be used as another alternative. When topping up note that about $\frac{1}{4}$ Imp pint (150cc) will raise the level from low to full.

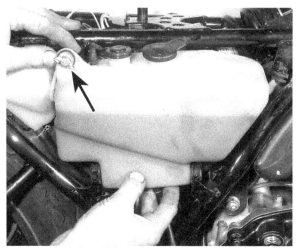
Release wing bolt and hinge oil tank outwards

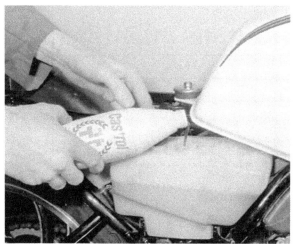
Top up using high quality two-stroke oil

Electrolyte level can be checked via transparent case

Coolant expansion tank is located behind oil tank

5 Control cable lubrication

Apply a few drops of motor oil to the exposed inner portion of each control cable. This will prevent drying-up of the cables between the more thorough lubrications that should be carried out during the 2000 mile/3 monthly service.

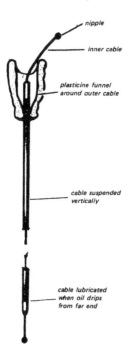

nipple

inner cable

plasticine funnel around outer cable

cable suspended vertically

cable lubricated when oil drips from far end

Control cable oiling

6 Rear chain lubrication and adjustment

In order that life of the rear chain can be extended as much as possible, regular lubrication and adjustment is essential.

The chain should be given a quick visual inspection once a week, and any accumulated road dirt removed with a petrol soaked rag. Lubricate the chain with one of the proprietory aerosol chain greases making sure that the lubricant reaches all sides of the chain. Ordinary engine oil is of limited value as a chain lubricant since most of it will be flung off the chain in use, but can be used in an emergency. Bear in mind that the excess

oil will probably lubricate the rear of the machine and the rider's back.

Check chain adjustment after lubrication. Place the machine on its centre stand and note the amount of up and down play at the middle of the lower run. Chains rarely wear evenly, so turn the wheel and re-check at intervals to establish the tightest point. The machine should now be allowed to rest on its wheels and the free play at the tightest point measured. This should be $1\frac{1}{4}$ – $1\frac{1}{2}$ in (30 – 40 mm).

If necessary, adjustment can be made after the wheel spindle nut and brake torque arm nut have been slackened. It may also prove necessary to slacken the rear brake adjuster. Tighten or slacken the drawbolt adjusters by an equal amount to preserve wheel alignment. This can be checked by way of the alignment marks on the fork ends. Once adjustment is complete, tighten the wheel spindle and torque arm and re-check the setting as described above. Note that the rear brake adjustment must be checked whenever the chain tension has been altered. Refer to Section 2 under the Monthly/1000 mile heading for details.

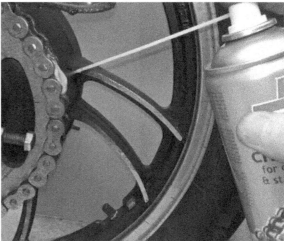
Aerosol lubricant is recommended for rear chain

Use index marks to preserve wheel alignment

7 Safety check

Give the machine a close visual inspection, checking for loose nuts and fittings, frayed control cables etc. Check the tyres for damage, especially splitting on the sidewalls. Remove any stones or other objects caught between the treads. This is particularly important on the rear tyre, where rapid deflation due to penetration of the inner tube will almost certainly cause total loss of control.

8 Legal check

Ensure that the lights, horns and trafficators function correctly, also the speedometer.

Master cylinder incorporates level window

Monthly or every 1000 miles (1500 km)

Complete the operations listed under the weekly/200 mile service heading, then carry out the following:

1 Checking the front brake

Check the operation of the front brake, looking for signs of sponginess which might indicate air in the system. Should sponginess or signs of fluid leakage be apparent, refer to Chapter 6 Sections 4, 5, 6 and 8 for more information, noting that it is imperative that any fault in the hydraulic system is repaired immediately.

The hydraulic fluid should be visible through the small inspection window in the rear face of the reservoir. If it has fallen to or below the minimum level line across the centre of the inspection window, remove the reservoir top and replenish the reservoir using a good quality hydraulic fluid meeting DOT 3 or SAE J1703 specifications. Under no circumstances should any other fluid or oil be used.

The hydraulic system is effectively sealed, and it is therefore unlikely that any appreciable loss of fluid will occur unless a leak has developed. If topping up proves necessary because of a sudden drop in fluid level, check the entire system for signs of leakage or seal failure. Pad wear will of course cause a drop in fluid level, but this will occur gradually.

Pad wear can be checked without any dismantling work being necessary. A groove runs down the centre of the friction material of each pad and denotes the maximum wear limit. If either pad is worn to or beyond this line both should be renewed. Yamaha recommend that the pad retaining pin, locking clip and shims should be renewed together with the pads, see Chapter 6 Section 3 for details.

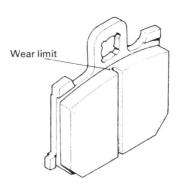

Wear limit

Front disc pad wear limit

2 Checking the rear brake

The rear brake is of the conventional single leading shoe drum type. It is simple in construction and operation, but unlike the front brake, does require regular adjustment to compensate for wear. The rear brake arm incorporates a pointer which is arranged to give an indication of the extent of wear of the brake shoes. If, with the brake pedal depressed, the pointer reaches the maximum wear line, it will be necessary to dismantle the brake and renew the shoes. See Chapter 6 Section 17 for details.

Before making any adjustment to the brake it is worth checking that the brake pedal height is set correctly. This is largely a matter of personal preference, the pedal being positioned so that it can be operated quickly and easily from the normal riding position. The normal setting is about 30 mm (1¼ in) below the top surface of the footrest, adjustment being effected by a stop bolt and locknut mounted at the front of the footrest mounting plate.

Once the pedal height has been checked and any necessary adjustment made, the brake pedal travel should be checked and adjusted. Again, the amount of travel is, to a point, a matter of choice, but as a guide 20 – 30 mm (¾ – 1¼ in) should be about right. Once adjusted, check that the brake does not drag when the pedal is released and that the brake light switch operates just as the brake begins to operate. The switch can be adjusted by turning the large plastic adjusting nut to alter the height of the switch.

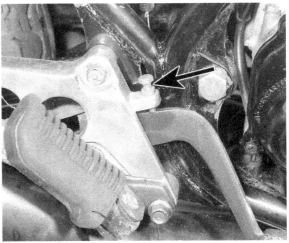

Brake pedal height is set by bolt and locknut

Brake adjustment is via nut on brake rod end

3 Lubricating the controls, cables and pivots

Clean and examine the various control lever and pedal pivots, lubricating each one with light machine oil or one of the multi-purpose maintenance aerosols. Check the outer cables for signs of damage, then examine the exposed portions of the inner cables. Any signs of kinking or fraying will indicate that renewal is required. To obtain maximum life and reliability from the cables they shold be thoroughly lubricated. To do the job properly and quickly use one of the hydraulic cable oilers available from most motorcycle shops. Free one end of the cable and assemble the cable oiler as described by the manufacturer's instructions. Operate the oiler until oil emerges from the lower end, indicating that the cable is lubricated throughout its length. This process will expel any dirt or moisture and will prevent its subsequent ingress.

If a cable oiler is not available, an alternative is to remove the cable from the machine. Hang the cable upright and make up a small funnel arrangement using plasticene or by taping a plastic bag around the upper end. Fill the funnel with oil and leave it overnight to drain through. Note that where nylon-lined cables are fitted, they should be used dry or lubricated with a silicone-based lubricant suitable for this application or a synthetic aerosol lubricant such as WD40 or similar. On no account use ordinary engine oil because this will cause the liner to swell, pinching the cable.

Check all pivots and control levers, cleaning and lubricating them to prevent wear or corrosion. Where necessary, dismantle and clean any moving part which may have become stiff in operation.

4 Checking the transmission oil

The transmission oil level is unlikely to drop between normal oil changes, but a regular monthly check will ensure that any sudden loss of lubricant is noticed before damage occurs. It is worth noting that loss of oil will be attributable to external leakage, which should be immediately obvious, or may be due to oil being drawn into the engine via worn crankshaft main bearing oil seals. Any such loss of oil should be investigated promptly.

Check the oil level by removing the filler plug and wiping off the small dip stick on its underside. Rest the plug back in its hole, but do not screw it home. Check that the oil level lies between the upper and lower level marks and replenish as required using SAE 10W/30 engine oil.

Three monthly or every 2000 miles (3000 km)

Complete the checks listed under the previous time/mileage headings and then carry out the following.

1 Checking the clutch adjustment

The correct clearance for the clutch assembly is set in two stages. Initial setting up is undertaken at the clutch release mechanism via a small screw and locknut housed under an inspection cover on the engine left-hand casing. Start by screwing fully home the cable adjuster at the handlebar end to give maximum free play in the cable. Remove the inspection cover at the rear of the left-hand casing and slacken the adjuster locknut. The adjusting screw should now be turned clockwise until a slight resistance is felt as it takes up any free play in the mechanism. Back off the adjuster by $\frac{1}{4}$ turn then, holding it in this position, secure the locknut.

The clutch cable adjuster should be set to give 2 – 3 mm (0.08 – 0.12 in) free play measured between the lever stock and arm. Release the knurled locking ring and move the adjuster in or out until the required clearance is obtained, then tighten the locking ring to secure the adjustment.

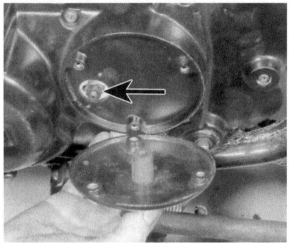

Remove inspection cover to reveal clutch adjuster

2 Carburettor adjustment

Throttle cable free play

The throttle cable adjuster must be set to give 3 – 7 mm (0.12 – 0.28 in) rotational free play measured at the inner flanged end of the twistgrip. The adjuster is located beneath the twistgrip housing and can be moved once the locknut has been slackened. Set the specified amount of clearance and operate the throttle a few times before final checking, then secure the locknut.

Synchronisation

It is important that the adjusters at the carburettor tops are set so that the two throttle valves are synchronised. If this is not the case, one cylinder will be trying to run faster than its neighbour and will in effect be 'carrying' it. This will result in uneven running and poor throttle response and fuel economy. To facilitate synchronisation, the two instruments each have a small window through which the throttle valve alignment marks are visible. Both can be viewed from the right-hand side of the machine. Open the throttle twistgrip fully and check that the two alignment marks coincide. If this is not the case, use the cable adjusters to bring them into synchronisation. Open and close the throttle twistgrip a few times to allow the cable and slides to settle, then re-check the adjustment.

Throttle stop and idle adjustment

This operation should only be carried out if the above two stages have failed to resolve a particular running problem. In most cases, poor synchronisation will account for the majority of cases of rough operation, but where the idle speed is incorrect or idling irregular, reference should be made to Chapter 3 Section 9 for further adjustment information. The specified idle speed with the engine at normal operating temperature is 1150 – 1250 rpm.

Synchronisation marks should be visible in windows

Use cable adjusters to synchronise throttle valves

3 Autolube pump adjustment

The engine depends upon the Autolube pump to deliver the required amount of fuel at any given throttle setting. It follows that it is important that the pump adjustment is maintained within the specified limits if it is wished to avoid the untimely demise of the engine. The pump settings should be checked **after** the throttle cable free play and carburettor synchronisation has been adjusted as described above.

Before any attempt at adjustment is made it is important to note the following remarks related to the pump alignment marks. On RD250 models before engine number RD250 4L1 100101 and RD350 models before engine number RD350 4L0 100101 the pump operating cable is set so that when the throttle twistgrip is fully open, the pump plunger pin is aligned with the raised index mark on the side of the pump pulley (Mark A).

In the case of machines fitted with later engines, check the identification number on the pump pulley. Where a 1M1 pulley is fitted, use the raised rectangular index mark as described above, but in the case of pumps fitted with a 4L1 pulley use the rectangular mark (Mark C) for 250 cc machines and the round mark for 350 cc models (Mark B).

Note: In the case of some earlier models, excessive oil consumption was a known problem. In some cases this can be corrected by making a new index mark 4 mm towards the weakest position, i.e. 4 mm towards the cable nipple, and using this mark for subsequent oil pump adjustment. In view of the risk of seizure involved if the pump is set to an excessively weak position, check first with a reputable Yamaha dealer that this modification is required.

Remove the front section of the right-hand engine casing to reveal the pump. Open the throttle twistgrip fully and note the relationship between the pump plunger pin and the appropriate index mark. If the two do not coincide alter the position of the pump cable adjuster until they are in alignment.

To check the minimum stroke adjustment, start the engine and allow it to idle. Observe the front end of the pump unit, where it will be noticed that the pump adjustment plate moves in and out. When the plate is out to its fullest extent, stop the engine and measure the gap between the plate and the raised boss of the pump pulley using feeler gauges. Do not force the feeler gauge into the gap – it should be a light sliding fit. Make a note of the reading, then repeat the procedure several times. The largest gap is indicative that the pump is at its minimum stroke position. If the pump is set up correctly, the gap found should be as shown in the specifications of Chapter 3, noting that some of the earlier machines may have been modified. If it is incorrect it will be necessary to remove the nut securing the

adjustment plate and to add or subtract shims as required. If necessary, these can be purchased from Yamaha dealers in thicknesses of 0.3 and 0.5 mm (0.0118 and 0.0197 in). In practice, the pump is unlikely to require frequent adjustment.

Finally, if there is any suspicion of air having found its way into the pump system it will be necessary for the system to be bled. Failure to do this can lead to oil starvation and possibly to engine seizure. It should be stressed that if the oil level in the tank has been allowed to drop too low and air has entered the oil pipe, bleeding must be carried out immediately. Start by removing the bleed screw from the pump body. The screw can be identified by its sealing washer. Allow the oil to run down through the feed pipe and note any air bubbles which emerge with it. When it is certain that all traces of air have been removed the screw can be refitted.

Next, start the engine and allow it to idle. To ensure that the distributor section of the pump and the delivery pipe are clear, pull the oil pump cable so that the pump stroke is at maximum. Allow the engine to run for about two minutes, noting that the excessive smoking that may result will soon clear when the machine is next used.

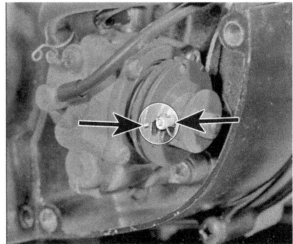

Alignment mark (see text) should coincide with plunger pin

Pump can be adjusted using cable adjuster

Check pump stroke adjustment with feeler gauges

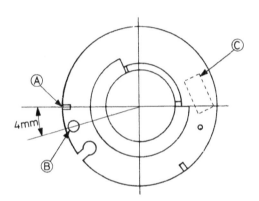

Oil pump pulley alignment marks

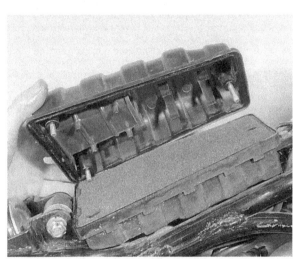

Remove air cleaner cover to expose element

4 Cleaning the air filter element

The air filter element on the Yamaha LC models is of the oiled foam type and is housed in a moulded plastic casing beneath the fuel tank. To gain access to the element the seat and tank must be removed as follows. Unlock the seat and lift it clear of the frame. Turn the fuel tap to the 'Off' position and pull off the fuel pipes. Remove the single rubber mounted bolt at the rear of the tank. Raise the rear of the tank slightly and then ease the tank backwards to disengage it from its mounting rubbers at the steering head.

Remove the air filter cover by unscrewing its three retaining screws. The flat foam element can now be lifted away. Wash the element in clean petrol to remove the old oil and any dust which has been trapped. When it is clean, wrap it in some clean rag and gently squeeze out the remaining petrol. The element should now be left for a while to allow any residual petrol to evaporate. Soak the cleaned element in engine oil and then squeeze out any excess to leave the foam damp but not dripping. Refit the element, ensuring that the cover seals properly. The tank and seat can now be refitted.

Note that a damaged element must be renewed immediately. Apart from the risk of damage from ingested dust, the holed filter will allow a much weaker mixture and may lead to overheating or seizure. It follows that the machine must never be used without the filter in position.

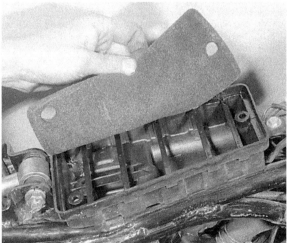

Air cleaner element can be removed for cleaning

5 Checking the steering and suspension

The handling and roadholding characteristics are to a great degree dependent upon the condition of the steering and suspension components. Although any sudden change will be immediately apparent to the rider, the gradual deterioration in performance caused by general wear can often be overlooked, and deficiencies may not be apparent unless the machine is ridden close to its limits. A regular and systematic check can, therefore, avoid unpleasant surprises on the road.

Roll the machine backwards and forwards with the front brake held on, noting the action of the front forks. These should operate smoothly and progressively and should show no signs of bouncing due to ineffective damping. Any signs of knocking felt through the handlebar or heard are often traced to slack or worn steering head bearings.

To check for wear, place the machine on its centre stand and place blocks or a crate beneath the crankcase so that the front wheel is raised clear of the ground. Persuade a friend to grasp the lower end of the fork legs and push and pull them. If play is present in the bearings it may prove easier to feel than see, by placing a finger between the frame headstock and the bearing shroud.

To adjust the steering head bearings, slacken the top bolt at the centre of the upper fork yoke. Using a C-spanner, slacken the adjustment nut by about a half turn, then slowly tighten it until a **slight** resistance is felt. It is very easy to overtighten the bearings, and this can cause stiff steering and rapid bearing wear. It is necessary to **just** eliminate free play and no more. When adjustment is complete check that the steering moves easily from lock to lock, then secure the top bolt to lock the adjustment. Any play remaining in the front end will probably be due to wear in the fork legs, and if found these should be overhauled as described in Chapter 5 Sections 2, 4 and 5.

To check for rear suspension wear, remove the blocks to allow the rear wheel to be raised off the ground. Grasp the rearmost end of the swinging arm assembly and push and pull it vigorously from side to side. Any discernible play indicates the need for pivot bearing overhaul, as described in Section 10 of Chapter 5.

The monoshock suspension unit can be adjusted for spring preload only, via a cam accessible when the seat and the small plastic tray have been removed. Using the C-spanner supplied in the toolkit, turn the cam ring to one of its five positions. Position 1 is the softest setting, 2 is the normal setting and 3, 4 and 5 give progressively harder settings for fast riding where handling is more important than comfort, or for passenger use.

There is no provision for damping adjustment, so if the rear wheel becomes uncharacteristically lively, have the unit checked by a Yamaha dealer. The units are built in Japan to DeCarbon designs, and whilst they should provide good service over a long period it is not unknown for them to expire prematurely. In the UK this has in the past meant a considerable delay whilst a new replacement is obtained from Japan, so if the unit seems to be wearing out it is advisable to order a new one as soon as is practicable.

6 Cleaning the exhaust system

The exhaust system on the LC models is of sophisticated design and must be kept clear of carbon accumulations to function efficiently. To this end, a detachable baffle is fitted in the end of each expansion chamber, retained by a single cross-head screw. Once the retaining screw has been released, free the baffle by grasping the bar in its end with a pair of pliers. This allows the baffle to be simultaneously twisted and pulled until it comes free.

Remove carbon deposits from the baffle by vigorous wire brushing. Internal deposits may be removed using a screwdriver or similar tool. When refitting the cleaned baffle it is advisable to use a copper-based grease on the screw threads to facilitate subsequent removal.

Rear suspension adjuster is located beneath seat

7 Spark plugs

Remove and examine each spark plug in turn, comparing its general condition with the colour photographs in Chapter 4 to gain an impression of general running conditions. The standard plug is an NGK B8ES, this being considered the best type for general use. If a lot of hard riding is done the standard plugs may show signs of overheating, with the porcelain insulator nose appearing bistered and the electrodes heavily eroded. Assuming that the ignition and carburation systems are set accurately, change the plug for the next coldest grade of that type, NGK B9ES or equivalent.

Conversely, if the machine is normally used for short trips or is ridden gently there may be some inclination towards plug fouling giving the plug a sooty or oily appearance. Again the fuel and ignition systems can affect this aspect, but if all is in order change to the next hottest grade, NGK B7ES or equivalent.

If the plug grade is correct it will now be necessary to clean and adjust the electrode gap. Note that if the electrodes show signs of excessive wear the plugs should be renewed, otherwise the opposing faces of the electrodes can be cleaned up and burnished using fine abrasive paper. It is inadvisable to leave abrasive particles on the plug end on case they find their way into the bores and cause scoring. To this end the plug should be washed in petrol to remove residual oil and the electrodes cleaned with the plug facing downwards.

Check the electrode gap using feeler gauges and if necessary adjust it to 0.7 – 0.8 mm (0.028 – 0.032 in) by bending the outer (earth) electrode. On no account attempt to bend the centre electrode – this will only result in a cracked insulator nose. Check that the plug threads are clean and wipe a trace of graphite or molybdenum grease on them to prevent their sticking in the cylinder head threads. Fit the plugs by hand, then tighten them just enough to obtain an effective seal. The specified torque figure is 2.0 kgf m (14.5 lbf ft).

8 Changing the transmission oil

With the engine at normal operating temperature to assist draining remove the drain plug on the underside of the crankcase and allow the oil to drain. A drain tray of about 4 pints (2 litre) capacity will be required. When the oil has drained, clean the plug and casing threads and refit the plug, tightening it to 2.0 kgf m (14.5 lbf ft). If there is any doubt as to the condition of the sealing washer, it should be renewed.

Remove the filler plug and add 1.5 litre (2.64 Imp pint) of SAE 10W/30 SE motor oil. Run the engine for a few minutes and then with it stopped, unscrew and clean the filler

Top up crankcase with SAE 10W/40 oil

Filler plug incorporates dipstick

plug/dipstick and place it on the top of the filler orifice so that the threads of the plug rest on the casing – do not screw the assembly home since this will give a false reading. Check that the oil level is between the minimum and maximum level marks, topping up as required.

9 Cleaning and lubricating the chain

Carry out the chain lubrication and adjustment sequence described under the Weekly/200 mile heading, Section 6, but remove the chain for more thorough cleaning and full lubrication as described below. It should be noted that although all models should be fitted with a conventional chain with a joining link it is possible that an endless chain may be encountered. This was the case with the machine featured throughout the manual and it was found that the chain passed through the rear sub-frame and thus could not be separated without a chain rivet extractor. It is recommended that, in the event of an endless chain being fitted, lubrication should be confined to the aplication of aerosol chain grease until renewal is required. A normal, removable, chain can then be fitted.

Locate the joining link and displace the spring clip using a pair of pliers. Prise off the side plate with a small screwdriver then slide the ends of the chain apart. Where possible connect a length of old chain to one end of the machine's chain and use this to run it off the sprockets. This will make installation much easier.

Place the chain in an old bowl fitted with paraffin or degreasing solvent and scrub it clean with a wire or stiff bristle brush. Allow the paraffin to drain and then give the chain a final rinse in clean petrol. Hang it up and allow any residual petrol to evaporate.

The cleaned chain can now be lubricated by immersing it in hot chain lubricant such as Linklyfe or Chainguard, following the manufacturer's directions. It is best not to carry out this operation in the kitchen since the smell of hot chain grease and accidental drips or splashes may not be appreciated by the kitchen's normal users. A small camping stove is ideal, but if used in the garage or workshop keep well clear of any flammable liquids. Do not overheat the lubricant: apart from increasing the danger of scalds from hot grease the chain hardening may suffer.

The grease will be almost watery when hot and so will penetrate between the rollers, bushes and pins, thus ensuring full lubrication of all moving parts. The chain can be removed when well soaked and hung over the tin of chain grease until cool.

Refit and adjust the chain, noting that readjustment may be required after a few days when any excess grease has been displaced.

10 General lubrication and maintenance

Give the machine a close visual examination looking for loose fittings or fasteners, tightening them as required. Clean and lubricate the control and stand pivots. Check the frame and cycle parts for signs of corrosion or paint damage and repair or repaint as required.

Six monthly or every 4000 miles (6000 km)

Complete the tasks listed under the previous service headings and then carry out the following:

1 Decarbonising the engine

Decarbonisation, or 'decoking' as it is often called, is an essential periodic maintenance task on all two-stroke engines. Because oil is deliberately burnt during combustion, the rate of carbon build up is greater than in a four-stroke engine and in time performance is impaired. If the carbon is not removed it may ultimately build up to the point where localised heating during combustion causes seizure or a holed piston.

The decarbonisation operation is rather more complicated on the LC models than on conventional air-cooled two-strokes because it is necessary to drain the cooling system. Although Yamaha recommend that it is carried out at 4000 mile intervals it has been found in practice that these models run cleaner than many two-strokes given that excessive short town journeys are avoided. It may be advisable to reduce the frequency of decarbonisation somewhat in view of the amount of work involved. It should also be noted that many Yamaha Dealers feel that this job should only be undertaken by experienced mechanics, so novices may be better advised to leave this work to the local Yamaha Dealer. For the more ambitious, proceed as follows.

Before starting work make sure that new cylinder head, cylinder base and exhaust port gaskets are to hand. It is also worth checking the hoses and purchasing replacements for any that appear perished or cracked. A large bowl or bucket will be needed to catch the coolant when this is drained, but note that if this is close to two years old it should be discarded and fresh coolant added during reassembly.

Place the machine on its centre stand, leaving adequate room to work on both sides. Slacken and remove the screws which retain the right-hand side of the radiator shroud to the radiator. Lift the shroud away so that the radiator cap can be removed. Check that the engine is quite cold, then slowly remove the cap.

It will now be necessary to drain the system via the two cylinder barrel drain bolts. Obtain a plastic bowl or bucket to catch the coolant and make up some sort of chute to keep the coolant off the engine casings. A piece of stiff card will suffice for this purpose. Remove each of the cylinder drain plugs in turn and allow the water to drain thoroughly, then disconnect the radiator bottom hose and drain off any residual coolant. It is important to ensure that none remains in the system so that there is no risk of water finding its way into the engine when the cylinder heads are removed.

Pull off the temperature gauge sender lead and disconnect the radiator top hose from the adaptor stub. Release the two bolts which retain the second coolant hose adaptor to the underside of the cylinder head. Slacken the spark plugs, but do not remove them completely. Release the carburettor assemblies from their mounting stubs, leaving the cables attached.

The cylinder head is retained by eight sleeve bolts, each being identified by a number cast into the cylinder head material. Slacken each one by $\frac{1}{2}$ turn, starting with No. 8 and working in numerical sequence to No. 1. When all have been slackened in this way the bolts can be removed completely. Before attempting to remove the cylinder head have some tissue or absorbent rag nearby to soak up any spilt coolant.

Break the seal between the cylinder barrels and the head by tapping around the latter with a soft-faced mallet or by judicious levering. Lift the cylinder head clear of the barrels and wipe up any drops of coolant, especially those which find their way into the bores. Remove any residual coolant by inverting and shaking the cylinder head.

Remove the carbon deposits from the inside of the cylinder heads using a hardwood or soft metal scraper to avoid scoring the soft aluminium alloy casting. Although not strictly essential it is a good idea to get the smoothest possible finish on the inside of the combustion chambers because this will improve gas flow and reduce the rate at which carbon builds up in the future. If it is felt worthwhile, some time spent with fine grades of abrasive paper and then metal polish can produce a mirror finish.

To clean the pistons and exhaust ports properly, the cylinder barrels should be removed. Partially remove one of the barrels until there is sufficient room to pack some clean rag around the connecting rod to catch any debris which might drop down as the piston emerges from the bore. Finish removing the barrel, supporting the piston as it emerges from its bore. Remove the remaining barrel, taking care to ensure that the crankcase mouths are adequately packed to avoid any risk of debris entering the engine.

Examine the pistons rings, piston and bores for signs of damage, taking remedial action as required (see Sections 17 and 18 of Chapter 1 for details). If all seems well, proceed as follows.

Carefully scrape away any carbon deposits from the exhaust ports, taking care not to burr the edge of the bore – any nicks at this point could catch and break the piston rings. Try to prevent any of the carbon debris from entering the inlet or transfer ports. This is best accomplished by packing the cylinder bore with clean rag.

Remove the carbon from the piston crowns using a hardwood or soft metal scraper. Like the cylinder head, the pistons will benefit from a polished finish, but this is by no means essential. Check that the piston rings are free in their grooves, and if necessary remove the rings and clean them. When decarbonisation is complete, clean the top of the crankcase and the rag around the connecting rods with a vacuum cleaner.

When reassembling the engine, make sure that the gasket faces are clean and dry and always use new gaskets. If the old gaskets are re-used it is likely that leakage will develop, and in some cases this will allow coolant to be forced into the crankcase. Not surprisingly, the main and big-end bearings will not respond well to water cooling and will soon corrode and wear out. The pistons should be coated with oil prior to insertion in their bores. Check that the ring ends are properly located by the pins in the ring grooves, then feed each one into the bore by hand. A tapered lead in makes this operation

reasonably easy.

Once the rings are engaged in the bore, remove the rag from the crankcase mouths. It is important that the barrels are slid into place squarely. This is because the piston ring ends will tend to catch in the square notch in the inlet passage if the barrel is slightly twisted (see photographs 34.4a and 34.4b in Chapter 1).

Place the new cylinder head gasket in position, followed by the cylinder head itself. The sleeve bolts should be fitted and tightened in numerical sequence. Tighten the bolts in two stages, the final torque figure being 2.4 kgf m (17.4 lbf ft).

Complete reassembly and refill the cooling system with 50% ethylene glycol antifreeze and 50% distilled water. Refer to item 4 in the weekly/200 Routine Maintenance schedule for further removals concerning the coolant specifications.

2 Overhauling the cooling system

The cooling system should be checked for signs of leakage or damage and any suspect hoses renewed. This is best undertaken in conjunction with decarbonisation when many of the parts concerned will have been removed. The used coolant can be put back into the system if it is kept clean, but should be renewed every two years as a matter of course. For further information, refer to Chapter 2.

3 Lights and electrical components

Examine the lighting system and direction indicators, renewing any cracked or damaged lenses and blackened or blown bulbs. Examine the wiring, looking for and repairing any chafed or cut insulation. Ensure that the various connectors are intact, and spray them with WD40 or similar to displace any moisture. Check the operation of all switches, cleaning and lubricating them with WD40. If any electrical faults are found, refer to Chapter 7 and to the wiring diagrams at the back of the manual.

4 Checking the ignition timing

To check the timing with the required degree of accuracy it will be necessary to use a dial gauge and a stroboscopic timing lamp. A dial gauge can be purchased through Yamaha Dealers as a set comprising the gauge, an adaptor and an extension needle as part number 90890-01252. A stroboscopic timing lamp, or 'strobe' is also available as part number 90890-03109. Both items can also be purchased from independent suppliers, many of whom advertise regularly in the motorcycle press. These tools are essential, and if they are not available the work must be entrusted to a Yamaha dealer.

Start by removing the left-hand engine casing to expose the alternator rotor. This will necessitate removal of the gearchange linkage. Remove the left-hand spark plug and screw the dial gauge adaptor into the spark plug hole. Fit the dial gauge extension needle to the gauge, then fit the assembly into the adaptor. Rotate the crankshaft by turning the alternator rotor. As the piston approaches top dead centre (TDC) the gauge reading will increase, stopping momentarily as TDC is reached and then decreasing as the piston begins to descend. Rock the crankshaft to and fro until the exact position of TDC is found, then set the gauge to read zero at this point. Move the rotor back and forth a few times to make sure that the needle does not move past zero.

Observing the gauge, turn the rotor clockwise until a reading of 3 or 4 mm is shown, then slowly move it anticlockwise until the piston is 2.0 mm (0.08 in) BTDC. Check the 'F' mark on the rotor in relation to the fixed index mark on the baseplate extension. If necessary, slacken the three baseplate bolts and move the baseplate until the marks coincide. The bolts can be slackened using an open ended spanner passed between the baseplate and the rotor edge. When the timing is correct, tighten the bolts and re-check the setting as described above.

Remove the dial gauge assembly and refit the spark plug. Connect the timing light as directed by the manufacturer's instructions, then start the engine. Direct the light at the 'F' mark on the rotor. The timing lamp will flash each time the plug sparks and will make the timing mark appear to freeze at that point. If all is well, the 'F' mark should coincide with the fixed index mark at 2000 rpm. In the unlikely event that there is a discrepancy, note where the 'F' mark does appear, stop the engine, and inscribe a new fixed index line in the appropriate place on the fixed stator. The new line should be used for subsequent timing checks.

Yearly, or every 8000 miles (12 000 km)

The items listed under this heading constitute a yearly overhaul which should be done in conjunction with the tasks listed under the previous service interval headings:

1 Overhauling the carburettor

If it is felt necessary, remove and overhaul each carburettor as described in Chapter 3. Note that unless the machine is actually running badly dismantling should be avoided and attention to the instruments confined to idle and pilot mixture adjustments and to checking synchronisation as described under the Three monthly/2000 mile heading.

2 Checking wheel and tyre condition

Check the condition of the wheels and tyres, not forgetting the wheel bearings which should be examined for wear and re-packed with grease. Check that the front wheel is correctly balanced and that both wheels are within run-out service limits. See Chapter 6 for details.

3 Changing the hydraulic fluid

Yamaha recommend that the hydraulic fluid in the front disc brake system should be changed annually, or every 4000 miles/6000 km. This is accomplished by pumping out the old fluid and then refilling and bleeding the system. Refer to Chapter 6 for details.

4 Changing the fork oil

Yamaha recommend that the damping oil in the front forks is changed every 4000 miles/6000 km. Unfortunately, no drain plug is provided so it will be necessary to remove the fork legs so that they can be inverted to remove the old oil. This procedure is described in Chapter 5 Section 2. The standard

Check that F mark aligns with index mark

fork oil grade is SAE 10W/30, but a lighter or heavier grade of oil may be used to alter the damping characteristics to suit riding conditions and personal preference. Bel-Ray produce a range of high quality fork oils for this purpose.

5 General greasing

The items listed below should be removed or dismantled for cleaning and examination, and re-packed with the appropriate grease. In most instances a fair amount of dismantling work must be undertaken, so it is recommended that the appropriate overhaul sequence is studied before starting work.

The following items should be lubricated with a good quality multi-purpose high melting-point lithium grease, such as Castrol LM or equivalent.

1   Steering head bearings
2   Wheel bearings and seals
3   Speedometer drive gearbox
4   Swinging arm pivot
5   Stand pivots
6   Brake and gearchange pivots
7   Rear brake cam

Castrol Lubricants

Castrol Engine Oils
Castrol Grand Prix

Castrol Grand Prix 10W/40 four stroke motorcycle oil is a superior quality lubricant designed for air or water cooled four stroke motorcycle engines, operating under all conditions.

Castrol Super TT Two Stroke Oil

Castrol Super TT Two Stroke Oil is a superior quality lubricant specially formulated for high powered Two Stroke engines. It is readily miscible with fuel and contains selective modern additives to provide excellent protection against deposit induced pre-ignition, high temperature ring sticking and scuffing, wear and corrosion.
Castrol Super TT Two Stroke Oil is recommended for use at petrol mixture ratios of up to 50:1.

Castrol R40

Castrol R40 is a castor-based lubricant specially designed for racing and high speed rallying, providing the ultimate in lubrication. Castrol R40 should never be mixed with mineral-based oils, and further additives are unnecessary and undesirable. A specialist oil for limited applications.

Castrol Gear Oils
Castrol Hypoy EP90

An SAE 90 mineral-based extreme pressure multi-purpose gear oil, primarily recommended for the lubrication of conventional hypoid differential units operating under moderate service conditions. Suitable also for some gearbox applications.

Castrol Hypoy Light EP 80W

A mineral-based extreme pressure multi-purpose gear oil with similar applications to Castrol Hypoy but an SAE rating of 80W and suitable where the average ambient temperatures are between 32°F and 10°F. Also recommended for manual transmissions where manufacturers specify an extreme pressure SAE 80 gear oil.

Castrol Hypoy B EP80 and B EP90

Are mineral-based extreme pressure multi-purpose gear oils with similar applications to Castrol Hypoy, operating in average ambient temperatures between 90°F and 32°F. The Castrol Hypoy B range provides added protection for gears operating under very stringent service conditions.

Castrol Greases

Castrol LM Grease

A multi-purpose high melting point lithium-based grease suitable for most automotive applications, including chassis and wheel bearing lubrication.

Castrol MS3 Grease

A high melting point lithium-based grease containing molybdenum disulphide. Suitable for heavy duty chassis application and some CV joints where a lithium-based grease is specified.

Castrol BNS Grease

A bentone-based non melting high temperature grease for ultra severe applications such as race and rally car front wheel bearings.

Other Castrol Products

Castrol Girling Universal Brake and Clutch Fluid

A special high performance brake and clutch fluid with an advanced vapour lock performance. It is the only fluid recommended by Girling Limited and surpasses the performance requirements of the current SAE J1703 Specification and the United States Federal Motor Vehicle Safety Standard No. 116 DOT 3 Specification.
In addition, Castrol Girling Universal Brake and Clutch fluid fully meets the requirements of the major vehicle manufacturers.

Castrol Fork Oil

A specially formulated fluid for the front forks of motorcycles, providing excellent damping and load carrying properties.

Castrol Chain Lubricant

A specially developed motorcycle chain lubricant containing non-drip, anti corrosion and water resistant additives which afford excellent penetration, lubrication and protection of exposed chains.

Castrol Everyman Oil

A light-bodied machine oil containing anti-corrosion additives for both household use and cycle lubrication.

Castrol DWF

A de-watering fluid which displaces moisture, lubricates and protects against corrosion of all metals. Innumerable uses in both car and home. Available in 400gm and 200gm aerosol cans.

Castrol Easing Fluid

A rust releasing fluid for corroded nuts, locks, hinges and all mechanical joints. Also available in 250ml tins.

Castrol Antifreeze

Contains anti-corrosion additives with ethylene glycol. Recommended for the cooling system of all petrol and diesel engines.

Chapter 1 Engine, clutch and gearbox

Contents

General description 1
Operations with the engine/gearbox unit in the frame 2
Operations with the engine/gearbox unit removed from
the frame 3
Removing the engine/gearbox unit from the frame 4
Dismantling the engine and gearbox: general 5
Dismantling the engine/gearbox unit: removing the
cylinder head, barrels and pistons 6
Dismantling the engine/gearbox unit: removing the
alternator 7
Dismantling the engine/gearbox unit: removing the
clutch 8
Dismantling the engine/gearbox unit: removing the
crankshaft and pump drive pinions 9
Dismantling the engine/gearbox unit: removing the
kickstart mechanism 10
Dismantling the engine/gearbox unit: separating the
crankcase halves 11
Dismantling the engine/gearbox unit: final dismantling 12
Examination and renovation: general 13
Gearbox input and output shafts: dismantling and
reassembly 14
Big-end and main bearings: examination and renovation ... 15
Oil seals: examination and renovation 16
Cylinder barrels: examination and renovation 17

Pistons and piston rings: examination and renovation 18
Cylinder head: examination and renovation 19
Gearbox components: examination and renovation 20
Kickstart mechanism: examination and renovation 21
Primary drive: examination and renovation 22
Clutch assembly: examination and renovation 23
Engine reassembly: general 24
Engine reassembly: refitting the tachometer drive 25
Engine reassembly: refitting the selector mechanism 26
Engine reassembly: refitting the gearbox components 27
Engine reassembly: refitting the crankshaft 28
Engine reassembly: joining the crankcase halves 29
Engine reassembly: fitting and adjusting the gear selector
shaft 30
Engine reassembly: fitting the kickstart mechanism idler
pinion and crankcase fittings 31
Engine reassembly: refitting the clutch, primary drive
and pump drive pinion 32
Engine reassembly: refitting the alternator, neutral switch
and left-hand outer cover 33
Engine reassembly: refitting the pistons, cylinder barrrels
and cylinder head 34
Fitting the engine/gearbox unit in the frame 35
Engine reassembly: final connections and adjustments 36
Starting and running the rebuilt engine 37

Specifications

Engine	RD250 LC	RD350 LC
Type	Water-cooled, parallel twin cylinder, two-stroke	
Bore	54 mm (2.126 in)	64.0 mm (2.520 in)
Stroke	54 mm (2.126 in)	54.0 mm (2.126 in)
Compression ratio	6.2 : 1	6.2 : 1
Capacity	247 cc (15.07 cu in)	347 cc (21.17 cu in)

Cylinder head		
Type	Cast aluminium alloy, one piece with integral water passages	
Maximum warpage	0.1 mm (0.004 in)	0.1 mm (0.004 in)
Head gasket thickness	1.2 mm (0.047 in)	1.2 mm (0.047 in)

Cylinder barrels		
Type	Cast aluminium alloy with cast-in iron sleeve	
Standard bore size	54.00 mm (2.126 in)	64.0 mm (2.520 in)
Service limit	54.10 mm (2.130 in)	64.10 mm (2.524 in)
Maximum taper	0.05 mm (0.020 in)	0.05 mm (0.020 in)
Maximum ovality	0.01 mm (0.0004 in)	0.01 mm (0.0004 in)

Pistons and rings

Type	Light aluminium alloy with two keystone compression rings
Oversizes available	+0.25, 0.50, 0.75 and 1.0 mm (+0.010, 0.020, 0.030 and 0.040 in)
Piston/bore clearance:	
RD250 (1980)	0.050 - 0.055 mm (0.0020 - 0.0022 in)
RD250 (1981) and RD350	0.065 - 0.070 mm (0.0026 - 0.0028 in)
Ring end gap:	
Top ring	0.30 - 0.45 mm (0.012 - 0.018 in)
2nd ring	0.30 - 0.50 mm (0.012 - 0.020 in)
Ring/groove clearance:	
Top ring	0.02 - 0.06 mm (0.0008 - 0.0024 in)
2nd ring	0.03 - 0.07 mm (0.0012 - 0.0028 in)

Crankshaft assembly

Big-end bearing deflection	0.36 - 0.98 mm (0.014 - 0.039 in)
Big-end bearing axial clearance	0.25 - 0.75 mm (0.010 - 0.030 in)
Crankshaft runout	0.05 mm (0.002 in)

Clutch

Type	Wet, multi-plate
No of plain plates	6
No of friction plates	7
No of springs	6
Friction plate thickness	3.0 mm (0.118 in)
Service limit	2.7 mm (0.106 in)
Plain plate warpage limit	0.05 mm (0.002 in)
Clutch spring free length	34.9 mm (1.374 in)
Service limit	33.9 mm (0.008 in)

Primary drive

Type	Helical gear
Reduction ratio	2.87 : 1 (66/23T)

Secondary drive

Type	Chain
Reduction ratio:	
RD250 LC	2.563 : 1 (41/16T)
RD350 LC	2.438 : 1 (39/16T)

Gearbox

Type	6-speed, constant mesh
Ratios:	
1st gear	2.571 : 1 (36/14T)
2nd gear	1.778 : 1 (32/18T)
3rd gear	1.318 : 1 (29/22T)
4th gear	1.083 : 1 (26/24T)
5th gear	0.962 : 1 (25/26T)
6th gear	0.889 : 1 (24/21T)

Torque settings

	kgf m	lbf ft
Cylinder head bolts	2.4	17.0
Primary drive gear	6.5	46.0
Clutch centre nut	7.5	54.0
Clutch spring bolts	1.0	7.0
Gearbox sprocket	7.5	54.0
Kickstart lever	2.5	18.0
Gearchange lever	1.5	10.0
Reed valve	0.1	0.7
Alternator rotor	8.0	58.0
Exhaust pipe	2.4	16.0
Oil pump	0.4	3.0
Transmission drain plug	2.0	14.0
Outer cover screws	1.0	7.0
Tachometer drive retainer	0.4	3.0
Selector drum cam	1.0	7.0
Detent lever	1.4	10.0

1 General description

The Yamaha RD250 and RD350 LC models employ a water-cooled, twin-cylinder, two-stroke engine built in unit with the primary drive, clutch and gearbox. The engine features a light alloy one-piece cylinder head which incorporates cast-in passages for the coolant. Separate light alloy cylinder barrels are fitted, each having an integral cast iron liner. Induction is controlled by a combination of piston porting and reed valves. A pressed-up crankshaft is used, carried on four caged ball bearings. The big-end and small-end bearings are of the needle roller type.

Primary drive is by gears to the wet multi-plate clutch mounted on the end of the gearbox input shaft. The gearbox is of the six-speed constant mesh type. Lubrication of the gearbox and primary drive components is by oil bath, whilst engine lubrication is direct injection via a metered pump driven off the crankshaft.

2 Operations with the engine/gearbox unit in the frame

The items listed below can be overhauled with the engine/gearbox unit in place. When a number of these operations need to be undertaken simultaneously it is usually worthwhile taking the unit out of the frame to gain better access and more comfortable working. Engine removal is fairly straightforward and can be expected to take about one hour.

    a) Cylinder head, barrels and pistons
    b) Clutch assembly and primary gear
    c) Oil pump
    d) Water pump
    e) Kickstart mechanism
    f) Ignition pickup
    g) Alternator assembly
    h) Gear selector mechanism (external components only)
    i) Final drive sprocket

3 Operations with the engine/gearbox unit removed from the frame

To gain access to the items listed below it is essential that the engine unit is removed from the frame and the crankcase halves separated:

    a) Crankshaft assembly
    b) Gearbox components
    c) Gear selector drum and forks

4 Removing the engine/gearbox unit from the frame

1 Before commencing any dismantling work it will be necessary to drain the transmission oil. This is best done whilst the engine is hot, so if time permits remove the drain plug and leave the oil to drain thoroughly and the machine to cool down. The drain plug is located on the underside of the crankcase unit, between the exhaust pipes. A bowl or drain tray of about ½ gallon or 2 litres capacity will be required. Let the oil drain completely, then clean the drain plug and examine the sealing washer. The latter should be renewed unless it is in good condition. Refit the drain plug, tightening it to 14.5 lbf ft (2.0 kgf m).
2 It will now be necessary to drain the cooling system, but note that this must be done when the engine has cooled down to avoid any risk of scalding when the radiator pressure cap is released. The radiator cap cannot be removed until the plastic guard has been removed. The latter has a raised lip which prevents the cap from being removed or tampered with during normal use. The guard is secured to the radiator by four cross-head screws and may be lifted away once these have been released.

3 The radiator cap should be unscrewed slowly to allow any residual pressure to escape, after which it can be removed completely. The cooling system can now be drained. A suitable container, such as a plastic bowl or bucket can be used to catch the coolant. If the coolant is fairly new (less than two years old) it can be re-used, in which case make sure that the container is clean. It will be noted that the cylinder drain plugs are situated well inboard of the engine casings. It is advisable to make up some sort of chute to direct the coolant into the drain container. This need not be elaborate — a piece of stiff card will suffice.
4 Remove each cylinder drain plug in turn and allow the coolant to run into the drain container. Slacken the clips which secure the top and bottom hoses to the radiator and displace the hose lower ends, remembering that some residual coolant will be released as the bottom hose is freed. Pull off the overflow hose from the radiator filler neck. The four bolts which secure the radiator can now be removed and the radiator removed to a safe place.
5 Check that the fuel tap is turned off, then prise off the fuel pipes having eased the clips away first. Release the seat latches and lift the seat away to expose the single bolt which secures the fuel tank. Remove the bolt and free the tank by lifting it and pulling it rearwards. The plastic side panels should also be removed by pulling them away from the frame, and placed with the tank in a safe place. Disconnect and remove the battery.
6 Dismantle and remove the exhaust system by releasing the two front nuts and the single rear bolt and nut which retain each half of the system. Move each half forward to free it from the retaining studs, then lift it clear of the machine.
7 Slacken the pinch bolt that retains the kickstart lever to its shaft. Remove the bolt and slide the lever off. Remove the front section of the engine right-hand casing, noting that it may prove necessary to use an impact driver to release the screws. The oil feed pipe from the tank is removed next, noting that some provision must be made to plug the open end of the pipe to stop the oil draining from the tank. A length of wooden dowel or a suitably sized bolt can be used as a plug, or alternatively, the tank can be drained into a can or bottle and the oil kept for re-use. Remove the pipe from the rubber fillet at the front of the crankcase and lodge it clear of the engine. Rotate the pump pulley to obtain maximum slack in the cable, then disengage the latter from the pulley. The cable can now be pulled clear, and it too should be positioned clear of the engine.
8 Unscrew both carburettor tops and withdraw the throttle valves. The assemblies can be left attached to the cables and positioned clear of the engine unit. The throttle valves are handed and need not be marked as a guide for reassembly. Slacken the hose clips which clamp the carburettors to the intake adaptors and to the plenum chamber adaptors. Displace the small oil delivery pipes from the carburettor bodies. The carburettor bodies can now be manoeuvred clear of the machine and placed to one side. Free the tachometer drive cable by releasing the knurled ring which retains it.
9 Release the gearchange linkage at the engine end by removing the clamp bolt. It is not essential that the linkage is removed completely, but this is advisable to permit lubrication during reassembly. The pedal end of the linkage is secured to its pivot by a circlip and plain washer.
10 Slacken the clutch cable adjuster at the lever end to obtain maximum free play in the cable. Remove the seven screws which retain the engine left-hand casing to the crankcase and lift the casing away to gain access to the lower end of the clutch cable. Bend the security tang clear of the cable, which can now be displaced and removed.
11 Bend back and flatten the tab washer which secures the gearbox sprocket nut. Lock the final drive by applying the rear brake, then slacken and remove the nut followed by the tab washer. Slide the sprocket off the output shaft splines and disengage it from the chain. If this latter operation proves difficult, slacken the rear wheel spindle nut and the chain tensioners, and push the wheel forward to obtain the necessary chain slack.
12 Trace the alternator output leads back to the row of block

connectors below the dualseat. Displace the three relevant connectors from their mounting clips and separate them. Disengage the alternator wiring from the frame, noting that the harness passes through a guide loop on the rear mudguard. Coil the harness and place it on top of the crankcase. Remove the spark plug caps and the water temperature sender lead and lodge these clear of the engine. The unit is now ready for removal.

13 The removal operation really requires two people to avoid damage or complications due to snagged cables or wiring. It is just possible to remove the unit single-handed, but this course of action is not recommended. The engine/gearbox unit is mounted at two points by long through bolts carried in rubber mountings. The ends of the mounting bolts pass through frame lugs, those on the right-hand side of the frame being bolted into place. Remove the nuts from the through bolts, then tap the bolts through and remove them. It may help to support the engine unit so that its weight is taken off the bolts. In the case of the later models fitted with engine stabiliser bars beneath the crankcase, these should be removed and placed to one side.

14 Release the frame lugs by removing the small bolts which secure each one to the frame. The unit can now be lifted out from the right-hand side, taking care not to damage the frame paintwork in the process. Place the engine on a workbench ready for further dismantling.

4.2 Radiator shroud is secured by four screws

4.4a Remove drain plug and allow coolant to drain

4.4b Slacken top hose clip and release hose

4.4c Remove bottom hose and catch residual coolant

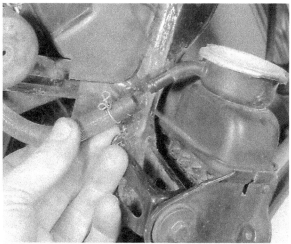

4.4d Pull off the radiator overflow pipe

4.4e Remove the four radiator mounting bolts ...

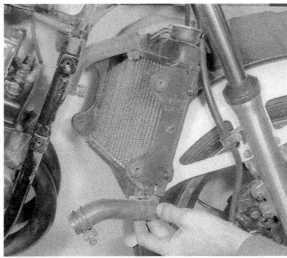

4.4f ... and lift the radiator clear of the frame

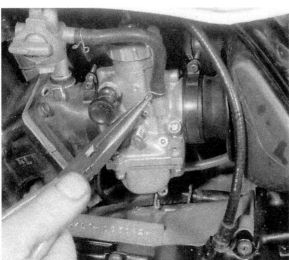

4.5a Prise off the petrol feed pipes at carburettors

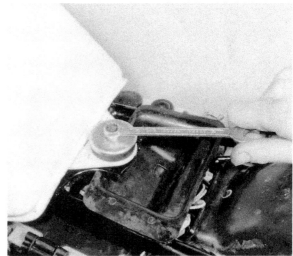

4.5b Remove single mounting bolt and lift tank away

4.5c Disconnect battery leads and release screw ...

4.5d ... to allow the battery to be removed

4.6a Exhaust pipe flange is retained by two nuts

4.6b Silencer is secured by a single bolt

4.7a Remove kickstart pinch bolt and slide off splines

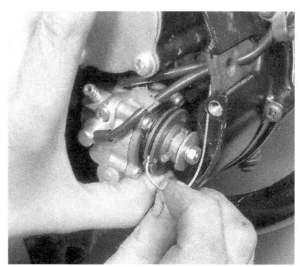

4.7b Disengage oil pump cable and oil pipes

4.8a Unscrew the carburettor tops and remove the valves

4.8b Slacken clips and disengage carburettors from stubs

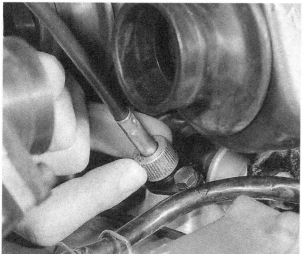

4.8c Unscrew knurled ring to free tachometer cable

4.9a Release clamp bolt to free gearchange linkage

4.9b Pedal is retained by a circlip

4.10a Remove left-hand outer cover ...

4.10b ... and disengage clutch cable

4.11 Remove nut and displace gearbox sprocket

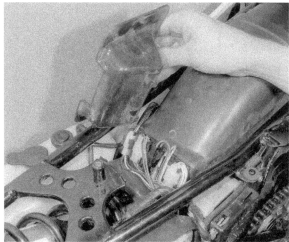

4.12 Electrical connectors are covered by plastic tray

4.13a Dismantle and remove front mounting ...

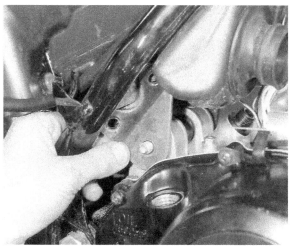

4.13b ... and rear mountings as shown

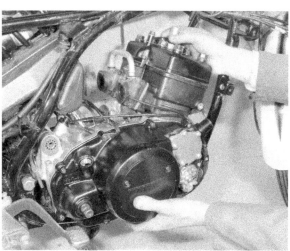

4.14 Engine unit can be lifted out to the right

5 Dismantling the engine and gearbox: general

1   Before commencing work on the engine unit, the external surfaces must be cleaned thoroughly. A motor cycle engine has very little protection from road grit and other foreign matter, which will sooner or later find its way into the dismantled engine if this simple precaution is not observed.
2   One of the proprietary engine cleaning compounds such as 'Gunk' or 'Jizer' can be used to good effect, especially if the compound is worked into the film of oil and grease before it is washed away. When washing down, make sure that water cannot enter the inlet or exhaust ports or the electrical system, particularly if these parts are now more exposed.
3   Never use force to remove any stubborn part, unless mention is made of this requirement in the text. There is invariably good reason why a part is difficult to remove, often because the dismantling operation has been tackled in the wrong sequence.
4   There are a number of special service tools available for the LC models, and whilst these are very useful it is possible to make do without them. The only possible exception is the flywheel rotor puller, Part Number 90890-01189. In the case of the machine featured in this manual it proved possible to use a

conventional three-legged puller as an alternative, but only if the rotor is not too firmly caught on the crankshaft taper. In view of the risk of damage that could be caused the author strongly recommends that the correct tool is purchased or borrowed – it costs far less than a new crankshaft or alternator.

6 Dismantling the engine/gearbox unit: removing the cylinder head, barrels and pistons

1   This operation can be undertaken with the engine unit in or out of the frame. In the former case it will first be necessary to drain the cooling system and to disconnect the carburettors, exhaust system, spark plug caps and water temperature sender lead. The whole dismantling sequence is described in Routine Maintenance in the section relating to decarbonisation.
2   When the cylinder head and barrels are removed care must be taken to prevent any residual coolant from finding its way into the engine. To this end, lift the above components away carefully, keeping them level to avoid spillage, and have some absorbent cloth or paper to hand so that any spillage can be mopped up.
3   Detach the radiator hose union from the cylinder head top

after removing the three bolts. The cylinder head is retained by eight sleeve bolts, each of which is marked by a number cast into the cylinder head material. The bolts should be slackened by a fraction of a turn at a time in the reverse of the numerical sequence, eg starting with bolt number 8 and working back to bolt number 1. Before the cylinder head can be lifted away it will be necessary to release the hose which runs between the right-hand crankcase and the cylinder head. This terminates in an adaptor at the upper end and is best removed with the adaptor by removing the two retaining screws. The hose is secured by a clip at the lower end.

4   The cylinder head can now be lifted away. If it proves to be stuck to the cylinder barrels, tap around the joint face with a hide or rubber mallet to break the seal. If necessary some degree of judicious levering can be employed using a steel ruler or a similar tool. There is plenty of room between the outside edge of the cylinder head and barrels and the gasket faces of each to permit this without risk of damage. Once the joint is broken, lift the head clear, wiping up any spilt coolant, especially where this has entered the cylinder bores.

5   The individual cylinder barrels should be removed in stages. Start by lifting the barrel slightly so that the crankcase mouth is exposed. Pack some clean rag around the connecting rod so that any debris which may drop as the piston emerges from the bore is prevented from entering the crankcase. Once the rag is

in place, remove the barrel and place it to one side. Repeat the operation with the remaining cylinder.

6   The pistons should be removed with the rag in position in case a circlip is dropped. Remove the outer circlip from one of the pistons by prising it free with an electrical screwdriver or by pulling it out with a pair of pointed-nose pliers. Displace the gudgeon pin by pushing it through from the opposite side until it clears the small-end eye and the piston can be lifted clear. If the pin is a tight fit, it may be necessary to warm the piston so that the grip on the gudgeon pin is released. A rag soaked in warm water will suffice, if it is placed on the piston crown. The piston may be lifted from the connecting rod once the gudgeon pin is clear of the small-end eye.

7   If the gudgeon pin is still a tight fit after warming the piston it can be lightly tapped out of position with a hammer and soft metal drift. **Do not** use excess force and make sure the connecting rod is supported during this operation, or there is a risk of it bending.

8   When the piston is free of the connecting rod remove the gudgeon pin completely, by removing the second circlip. Place the piston, rings and gudgeon pin aside for further attention, but discard the circlips. They should never be re-used; new circlips must be obtained and fitted during rebuilding. Remove the small-end bearing and place it inside the piston for safe keeping. Repeat the removal sequence on the remaining piston.

6.3a Slacken head bolts by reversing tightening sequence

6.3b Release hose adaptor from underside of head

6.6a Use pointed-nose pliers or screwdriver to free circlips

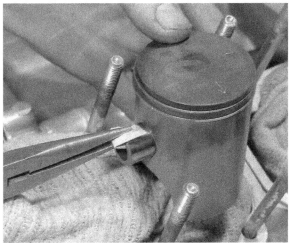

6.6b Displace gudgeon pin, using pliers where necessary

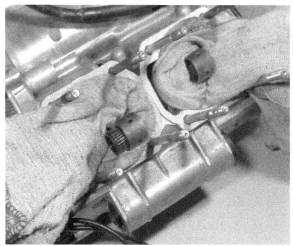

6.8 Small-end bearing can be pushed out

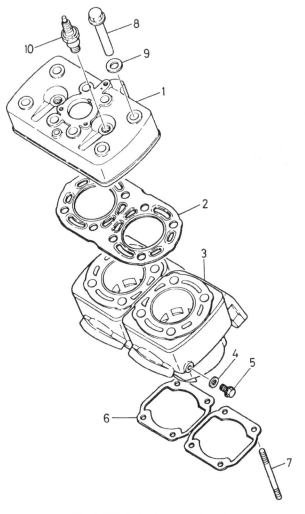

7 Dismantling the engine/gearbox unit: removing the alternator

1   The alternator assembly can be removed with the engine unit on the workbench or installed in the frame. In the latter case it will first be necessary to remove the gearchange linkage and the engine left-hand casing. Note that it is strongly recommended that the appropriate Yamaha rotor puller, part number 90890-01189, is obtained before work commences.
2   The crankshaft must be prevented from turning whilst the alternator rotor nut is slackened. If the engine is in the frame this can be accomplished by selecting top gear and locking the rear wheel by applying the brake. If the operation is being undertaken on the workbench a strap wrench can be fitted around the rotor periphery. An alternative method which can be used if the cylinder head, barrels and pistons have already been removed is to pass a smooth close-fitting metal bar through one of the connecting rod small-end eyes. The ends of the bar can then be supported on wooden blocks placed against the crankcase mouth. Once the crankshaft has been immobilised by one of the above methods, slacken and remove the securing nut.
3   The rotor is located on a taper, and this joint invariably becomes tight. It was found that in the case of the machine featured in the accompanying photographs a conventional legged puller could be used to draw the rotor off, but that the amount of pressure required could have caused damage if the rotor had been just slightly tighter. For this reason the above approach is not to be encouraged, but if it proves essential proceed as follows.
4   Assemble the puller as shown in the accompanying photograph, ensuring that the puller legs do not contact any part of the stator assembly. Screw down the centre bolt firmly but not excessively then tap the end of the bolt to jar the rotor free. If necessary, tighten the bolt a little more and tap the end of the bolt a few more times, but if this fails **do not** continue to apply further pressure or damage will almost certainly result. If necessary, abandon the attempt and take the unit to a Yamaha dealer who will be able to remove the rotor safely and easily.
5   If the correct tool is available, screw it into the extractor thread, then tighten the centre bolt to draw the rotor off. This method should dislodge the rotor easily, but if necessary the centre bolt can be tightened a little more and then struck again. This method will almost invariably succeed in removing the rotor, but if removal still proves difficult it should be noted that there will usually be a good reason for the rotor's reluctance to be separated from the crankshaft. It is suggested that such

Fig. 1.1 Cylinder head and barrel

1   Cylinder head
2   Cylinder head gasket
3   Cylinder barrels
4   Washer
5   Drain bolt
6   Cylinder base gasket
7   Stud – 8 off
8   Sleeve bolt – 8 off
9   Washer – 8 off
10 Spark plug

cases should be entrusted to a Yamaha dealer who will have the equipment, experience and sheer cunning necessary to complete safely the removal operation.
6   Before the stator is removed it is advisable to mark its position in relation to the crankcase by scratching a pair of alignment marks on the two components. This will ensure that the ignition timing is approximately correct, though this will need to be checked for accuracy when reassembly is complete. The stator is secured by three bolts which pass through elongated slots in the stator. Remove the bolts to free the stator then feed the wiring through the hole in the crankcase having displaced the grommet. Note that it will also be necessary to release the neutral switch lead.

7.2 Lock crankshaft and slacken rotor bolt

7.4 Assemble three-legged puller as shown

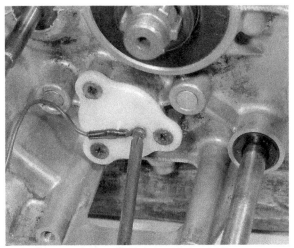

7.6a Release neutral switch lead from switch body

7.6b Note neutral switch lead grommet (arrowed)

7.6c Release mounting bolts and lift stator away

8 Dismantling the engine/gearbox unit: removing the clutch

1   The above mentioned parts may be removed with the engine in or out of the frame. In the former case it will be necessary to remove first the transmission drain plug and allow the oil to drain, and to release the kickstart lever and oil pump cable and pipes. Note that the water pump is housed in the right-hand engine casing and thus the cooling system must be drained before it can be removed, though it is not necessary to disturb either the water pump or the oil pump.

2   Release the screws around the outer edge of the engine casing. These are invariably tight and will require the use of an impact driver to loosen them without damaging the screw heads. The cover can now be lifted away complete with the pumps and placed to one side. If the work is being undertaken with the engine in the frame the nut which secures the primary drive pinion must be slackened at this stage if it is wished to remove the pinion. To prevent crankshaft rotation as the nut is removed, select top gear and apply the rear brake. Once the nut has been loosened the clutch can be dismantled. Where the engine is being stripped on the workbench it is easier to lock the crankshaft by passing a round metal bar through one of the connecting rod small-end eyes, its ends being supported on

small wooden blocks placed against the crankcase mouth to protect the gasket face.

3　Slacken and remove the six bolts which secure the clutch pressure plate, releasing them evenly by about one turn at a time until they are no longer under spring tension. Lift the pressure plate clear together with the six clutch springs. Displace and remove the clutch plain and friction plates.

4　It is now necessary to devise some method of holding the clutch centre whilst the retaining nut is slackened. Yamaha produce a clutch holding tool and this can be ordered through Yamaha dealers as part number 90890-01024. The tool is effectively a plain plate with a handle welded to it, and this can be improvised quite easily if an unwanted plate can be acquired. An alternative tool is shown in photo 32.3f, and was fabricated using a strip of mild steel plate. The photo shows the tool fitted for tightening the nut, but the principle is the same. One end hooks into the clutch centre splines whilst the free end passes through one of the clutch drum slots and rests against the crankshaft. Once locked, knock back the clutch centre nut locking washer and release the retaining nut. Great care should be taken to ensure that the holding tool does not slip; if it does damage is likely to occur. The clutch centre and drum can be slid off the gearbox input shaft together with the bush which carries the clutch drum and the thrust washer which fits between it and the clutch centre.

8.2 Remove screws and lift outer cover away

8.3a Remove clutch cover bolts and springs ...

8.3b ... followed by the pressure plate

8.3c The clutch plates can now be displaced

8.4a Home-made tool used to lock clutch centre (see photo 32.3f)

8.4b Remove clutch drum and centre bush

9 Dismantling the engine/gearbox unit: removing the crankshaft and pump drive pinions

1   The pinions referred to in the heading are mounted on the right-hand end of the crankshaft and are secured by a large nut. The crankshaft (primary drive) pinion is located by a Woodruff key, whilst the outer pinion, which drives the oil and water pumps, is located by pressure from the securing nut. It follows that the nut is very tight and will require the use of a stout socket and lever bar to facilitate loosening. A ring spanner can be used with good effect but on no account attempt removal with an open-ended spanner.

2   As mentioned in the previous Section, a secure method of holding the crankshaft is essential. Yamaha suggest that a wad of rag is jammed between the teeth of the primary drive pinion and the corresponding teeth of the clutch drum. This is a far from satisfactory approach and could lead to damage to the crankshaft or gearbox input shaft, and is thus not recommended by the author. If the engine is installed in the frame, it is recommended that the securing nut is slackened before the clutch is removed. This will allow the crankshaft to be locked through the transmission by selecting top gear and applying the rear brake.

3   If, on the other hand, the unit is to be dismantled on the workbench, wait until the cylinder head, barrels and pistons have been removed. A close-fitting round metal bar can now be passed through one of the connecting rod small-end eyes and its free ends supported by wooden blocks placed against the crankcase mouth. This will provide positive restraint for the crankshaft without risk of damage to any component. Slacken and remove the nut, followed by the Belville washer. The pump pinion can now be removed together with the primary drive pinion and its Woodruff key. If the clutch is to be dismantled, refer to Section 8.

10 Dismantling the engine/gearbox unit: removing the kickstart mechanism

1   The kickstart mechanism can be removed with the engine in or out of the frame after the engine right-hand casing has been detached. Note that if it is wished to remove the idler pinion which conveys drive to the clutch drum, and thus to the crankshaft, the clutch must be removed first. The idler pinion does not impair crankcase separation and may be left in position unless specific attention to it or the input shaft components is required.

2   Release the kickstart return spring by grasping its outer end with a pair of pliers and disengaging the end from its anchor pin. Allow the spring to unwind in a controlled manner, then pull the kickstart shaft assembly from its casing hole. If it is wished to remove the idler gear pinion after the clutch has been removed, release the circlip which retains it to the end of the gearbox output shaft.

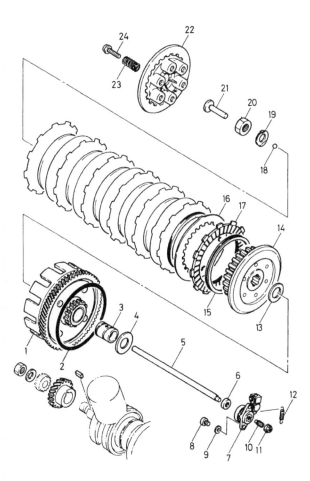

Fig. 1.2 Clutch

1 Clutch drum	13 Thrust washer
2 O-ring	14 Clutch centre
3 Bush	15 Damper ring – 7 off
4 Thrust washer	16 Plain plate – 6 off
5 Push rod	17 Friction plate – 7 off
6 Oil seal	18 Steel ball
7 Release mechanism	19 Locking washer
8 Screw – 2 off	20 Nut
9 Washer – 2 off	21 Push rod
10 Adjusting screw	22 Pressure plate
11 Locknut	23 Spring – 6 off
12 Spring	24 Bolt – 6 off

9.3a Lock crankshaft and slacken retaining nut

9.3b Remove the Belville washer ...

9.3c ... followed by the pump drive pinion ...

9.3d ... and the crankshaft pinion

10.2a Unhook the kickstart return spring from its anchor pin

10.2b Kickstart assembly can now be withdrawn

11 Dismantling the engine/gearbox unit: separating the crankcase halves

1   The remaining external engine components can be left in positon and do not impair crankcase separation. Their removal is described in Section 12, and it should be noted that if required they can be removed without complete dismantling of the unit. Bear in mind, however, that any internal components, such as the tachometer drive shaft cannot be dealt with unless the crankcase halves have been parted. The only remaining item to be removed at this stage is the input shaft right-hand bearing retainer, which bridges the crankcase halves. It is retained by two cross-head screws which are invariably stubborn and will require the use of an impact driver to effect safe removal.

2   The crankcase bolts are numbered to indicate the correct tightening sequence and should be released in reverse order, starting at the highest number and working backwards. Each bolt should be slackened initially by about $\frac{1}{4}$ turn, then removed completely. There are eight bolts on the underside of the unit and a further eight on the upper face of the crankcase. Once all of the bolts have been removed, separate the joint by striking the front and rear edges of the upper crankcase half with a soft-faced mallet.

3   When the joint has been broken the upper crankcase half can be lifted away. Note that the connecting rods will tend to fall against the crankcase edge and they should be supported to prevent this. The gearbox shafts and the crankshaft should remain in the lower crankcase half.

12 Dismantling the engine/gearbox unit: final dismantling

1   Grasp the ends of the crankshaft assembly and lift it away from the lower crankcase half. Note the half-ring which locates the right-hand main bearing. This will probably be displaced as the crankshaft is removed and should be placed in a safe place to avoid its loss. The gearbox input shaft and output shaft assemblies should be removed in a similar manner, again noting the locating half-rings.

2   Disengage the selector claw from the end of the selector drum and remove the gearchange shaft assembly by displacing the selector shaft on the opposite side of the crankcase. Note that the seal through which the shaft must pass is easily damaged and if it is necessary to re-use it, protect the seal lip by wrapping some pvc tape around the shaft splines.

3   Release the selector drum stopper arm by removing its single retaining bolt, then remove the selector drum retainer which is held by two countersunk crosshead screws. Note that the retainer also locates one of the selector fork shafts. The remaining shaft is retained by the selector mechanism centralising spring anchor pin which should be removed together with the retainer plate.

4   Working from inside the crankcase, use a pair of pointed-nose pliers to displace the circlips on the inner ends of the selector fork shafts whilst the shafts are pushed through the casing. Support the selector forks and withdraw the shafts completely, then slide the shafts back through the forks to keep them in the correct relative positions as a guide during reassembly. The selector drum and its bearing can now be pushed out of the casing and removed.

5   The tachometer drive need not be disturbed unless it requires specific attention, but if removal proves necessary proceed as follows. Remove the circlip and plain washer which retain the white plastic drive pinion, then remove the pinion from the shaft end. Displace the drive pin and place it with the pinion for safe keeping. Release the single bolt which retains the tachometer drive body to the crankcase and remove it complete with the driven shaft. Remove the single screw which retains the drive shaft locating plate to allow the shaft to be displaced and removed. The drive gear should be slid off the

11.1 Remove input shaft bearing retainer

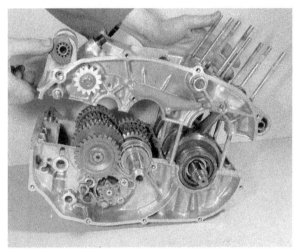

11.3 Release crankcase bolts and separate crankcase halves

shaft as the latter is pulled clear of the crankcase, having first released the circlips which retain it.

13 Examination and renovation: general

1   Before examining the parts of the dismantled engine unit for wear, it is essential that they should be cleaned thoroughly. Use a paraffin/petrol mix to remove all traces of old oil and sludge that may have accumulated within the engine.

2   Examine the crankcase castings for cracks or other signs of damage. If a crack is discovered, it will require professional repair.

3   Examine carefully each part to determine the extent of wear, checking with the tolerance figures listed in the main text or in the Specifications section of this Chapter. If there is any question of doubt, play safe and renew.

4   Use a clean, lint-free rag for cleaning and drying the various components. This will obviate the risk of small particles obstructing the internal oilways, causing the lubrication system to fail.

12.1 Remove gearbox shafts, noting position of half-rings

12.2 Displace gearchange shaft from crankcase

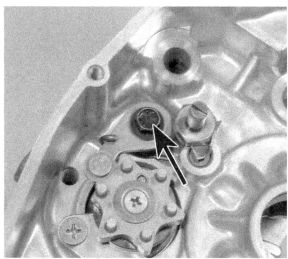

12.3a Remove pivot bolt and lift stopper arm away

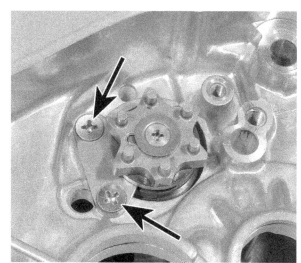

12.3b Release screws (arrowed) and remove drum retainer

12.3c Remove eccentric pin and selector fork shaft retainer

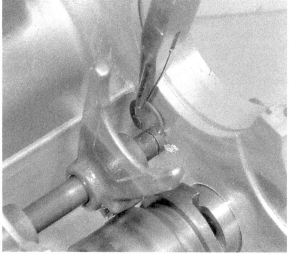

12.4a Use pointed-nose pliers to withdraw circlip ...

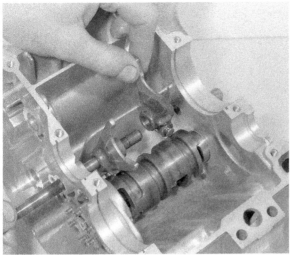

12.4b ... then slide selector fork shaft out of casing

12.4c Selector drum can now be removed as shown

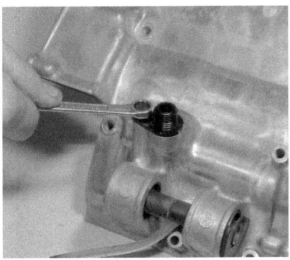

12.5a Remove single retaining bolt ...

12.5b ... and withdraw tachometer drive body

12.5c Release drive pinion and retainer

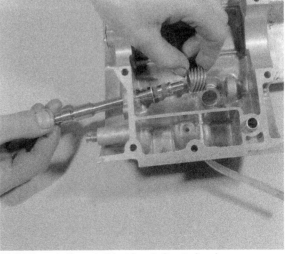

12.5d Free circlip and slide drive shaft out of casing

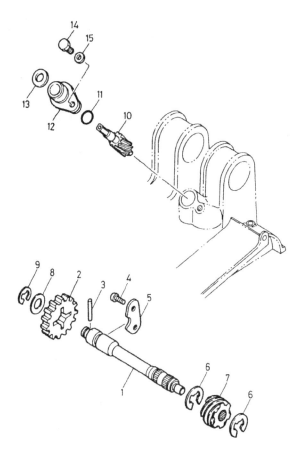

Fig. 1.3 Tachometer drive gear assembly

1   Tachometer drive shaft
2   Plastic drive pinion
3   Pin
4   Screw – 2 off
5   Locating plate
6   Circlip – 2 off
7   Drive gear
8   Washer
9   Circlip
10 Driven gear
11 O-ring
12 Drive body
13 Washer
14 Bolt
15 Washer

14 Gearbox input and output shafts: dismantling and reassembly

1   The gearbox clusters should not be disturbed needlessly, and need only be stripped where careful examination of the whole assembly fails to resolve the source of a problem, or where obvious damage, such as stripped or chipped teeth is discovered.

2   The input and output shaft components should be kept separate to avoid confusion during reassembly. Using circlip pliers, remove the circlip and plain washer which retains each part. As each item is removed, place it in order on a clean surface so that the reassembly sequence is self evident and the risk of parts being fitted the wrong way round or in the wrong sequence is avoided. Care should be exercised when removing circlips to avoid straining or bending them excessively. The clips must be opened just sufficiently to allow them to be slid off the shaft. Note that a loose or distorted circlip might fail in service, and any dubious items must be renewed as a precautionary measure. The same applies to worn or distorted thrust washers.

3   Having checked and renewed the gearbox components as required (see Section 20) reassemble each shaft, referring to the accompanying line drawing and photographs for guidance. The correct assembly sequence is detailed below.

Input shaft (mainshaft)

4   Note that the input shaft is readily identified by its integral 1st gear pinion. Slide the 5th gear into position with the dogs facing away from the 1st gear.

5   Fit the plain thrust washer and secure the 5th gear pinion with its circlip. The double 3rd/4th gear pinion is fitted next, with the smaller, 22 tooth, gear towards the 5th gear pinion. Fit a circlip to the next exposed groove, followed by a splined thrust washer. This retains the 3rd/4th gear pinion but allows it to move along the shaft to effect gear changes.

6   Slide the 6th gear pinion into place, noting that the engagement dogs face inwards, towards the 3rd/4th gear. The 2nd gear pinion is fitted next and is retained by a plain thrust washer and a circlip. The needle roller bearing should now be lubricated and slid into place to complete assembly. If it has been removed, fit the large caged ball bearing and large thrust washer to the right-hand end of the shaft.

Output shaft (layshaft)

7   Slide the 2nd gear pinion up against the shouldered portion of the output shaft, noting that it must be fitted from the right-hand end, with the engagement webs away from the shoulder. Fit a plain thrust washer and retain the pinion with a circlip.

8   Slide the 6th gear pinion into position with the selector groove away from the previous gear. Fit a circlip to limit the 6th gear pinion's movement then slide a splined thrust washer into place.

9   The 4th gear pinion is fitted next, noting that the heavily chamfered teeth face outwards, towards the right-hand end of the shaft. Secure it with a circlip, then fit the 3rd gear pinion, plain face inwards, and retain it with a splined thrust washer and a circlip.

10 The 5th gear pinion can now be slid into place with the selector groove inwards, followed by the large 1st gear pinion with its plain face outwards. Fit a plain thrust washer and a circlip to retain the above components.

11 Place the caged needle roller bearing over the right-hand end of the output shaft and the large ball bearing, seal and

spacer over the left-hand end. The idler gear which runs on the right-hand end of the output shaft can be fitted at this stage noting that a thrust washer is fitted on each side of the pinion and that the assembly is retained by a circlip.

12 For identification purposes the various gear pinions are listed below, together with their appropriate number of teeth. Note that this applies to both 250cc and 350cc models.

Input shaft	No of teeth
1st gear (integral with shaft)	14
2nd gear	18
3rd/4th gear	22/24
5th gear	26
Top gear	27

Output shaft	No of teeth
1st gear	36
2nd gear	32
3rd gear	29
4th gear	26
5th gear	25
Top gear	24

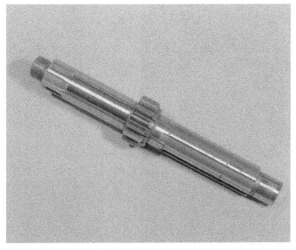

14.4a 1st gear pinion is integral with input shaft

14.4b Fit 5th gear pinion noting direction of dogs

14.5a Secure 5th gear pinion with plain thrust washer and circlip

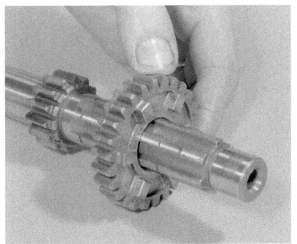

14.5b Fit 3rd/4th gear pinion facing in direction shown

14.5c Position circlip in next groove, then fit washer

14.6a Fit 6th gear pinion noting that dogs face inwards

14.6b Slide 2nd gear pinion into position ...

14.6c ... fit the plain thrust washer as shown ...

14.6d ... and retain with a circlip

14.6e The caged needle roller bearing can now be fitted

14.6f Slide large bearing onto opposite end of shaft ...

14.6g ... followed by large plain thrust washer

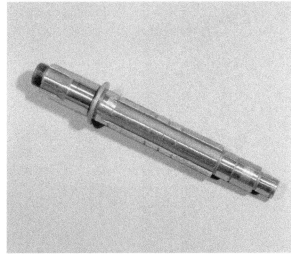

14.7a Output shaft has plain shoulder at sprocket end

14.7b Fit the 2nd gear pinion with engagement webs as shown

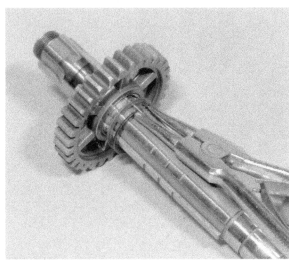

14.7c Slide thrust washer into position and retain with circlip

14.8a Fit 6th gear pinion, noting position of selector groove

14.8b Position circlip as shown above, then fit washer

14.9a Fit 4th gear pinion, noting direction of chamfered teeth

14.9b Retain 4th gear pinion with a circlip

14.9c Slide 3rd gear pinion into place. Fit splined washer ...

14.9d ... and secure with circlip

14.10a Fit 5th gear pinion, noting position of selector groove

14.10b Fit 1st gear pinion with plain face outwards

14.11a Fit the needle roller bearing to right-hand end ...

14.11b ... followed by the plain thrust washer ...

14.11c ... and idler pinion

14.11d Fit the final thrust washer ...

14.11e ... and retain the assembly with a circlip

14.11f Fit LH bearing with locating groove innermost

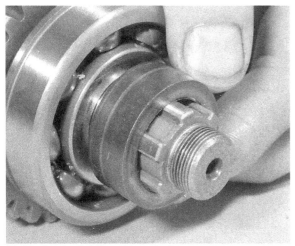

14.11g Bearing is followed by a steel spacer

14.11h Grease seal lip and fit it over spacer as shown

Fig. 1.4 Gearbox components

1 Output shaft
2 Bearing locating ring
3 Output shaft left-hand bearing
4 Oil seal
5 Spacer
6 Final drive sprocket
7 Tab washer
8 Nut
9 Output shaft 2nd gear pinion
10 Thrust washer
11 Circlip
12 Output shaft 6th gear pinion
13 Circlip
14 Splined thrust washer
15 Output shaft 4th gear pinion
16 Circlip
17 Output shaft 3rd gear pinion
18 Splined thrust washer
19 Circlip
20 Output shaft 5th gear pinion
21 Output shaft 1st gear pinion
22 Thrust washer
23 Circlip
24 Needle roller bearing
25 Bearing locating ring
26 Input shaft and 1st gear pinion
27 Input shaft right-hand bearing
28 Circlip
29 Bearing retainer
30 Screw – 2 off
31 Input shaft 5th gear pinion
32 Thrust washer
33 Circlip
34 Input shaft 3rd and 4th gear pinion
35 Circlip
36 Splined thrust washer
37 Input shaft 6th gear pinion
38 Input shaft 2nd gear pinion
39 Thrust washer
40 Circlip
41 Needle roller bearing

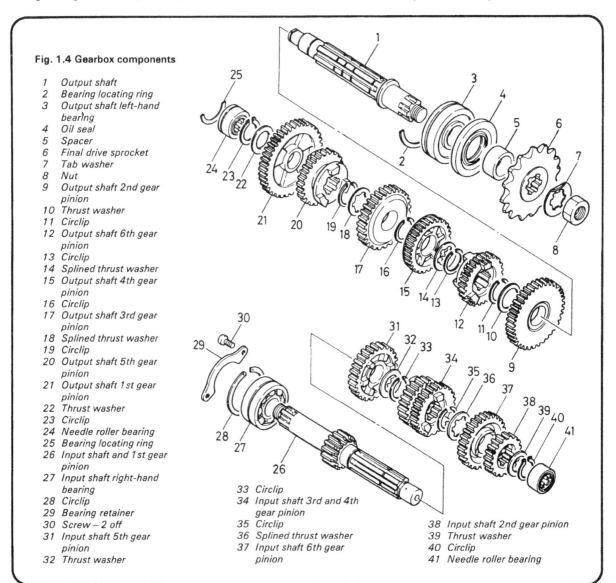

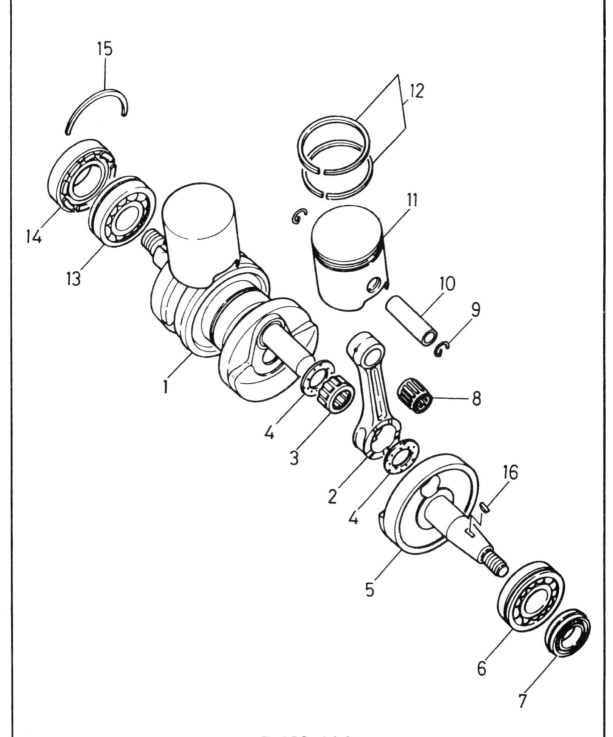

Fig. 1.5 Crankshaft

1 Crankshaft	7 Oil seal	12 Piston rings
2 Connecting rod	8 Small-end bearing	13 Main bearing
3 Big-end bearing	9 Circlip	14 Oil seal
4 Thrust washer	10 Gudgeon pin	15 Bearing locating ring
5 Flywheel	11 Piston	16 Woodruff key
6 Main bearing		

15 Big-end and main bearings: examination and renovation

1   Failure of the big-end bearing is invariably accompanied by a knock within the crankcase that progressively becomes worse. Some vibration will also be experienced.

2   There should be no vertical play whatsoever in the big-end bearings, after the oil has been washed out. If even a small amount of vertical play is evident, the bearings are due for replacement. (A small amount of endfloat is both necessary and acceptable). Do not continue to run the machine with worn big-end bearings, for there is risk of breaking the connecting rods or crankshaft.

3   If a dial gauge and V-blocks are available check the amount of radial clearance in the big-end bearings, measuring this as lateral deflection at the small-end of each connecting rod to magnify the clearance in the big-end bearings. A serviceable bearing will allow between 0.36 and 1.0 mm (0.0142 - 0.0394 in) free play, whilst 2.0 mm (0.0787 in) of movement is indicative of the need for renewal of the bearing concerned.

4   Check the connecting rod side clearance by measuring the gap between it and the adjacent flywheel boss with feeler gauges. Clearance should be between 0.25 mm (0.0098 in) minimum and 0.75 mm (0.0295 in) maximum.

5   The crankshaft main bearings are of the ball journal type. If wear is evident in the form of play, or if the bearings feel rough as they are rotated, replacement is necessary. Always check after the old oil has been washed out of the bearings. Whilst it is possible to remove the outer bearings at each end of the crankshaft, it is probable that the centre bearing will also require attention.

6   In the event that the big-end or main bearings prove to be in need of renewal it will be necessary to have the work done by an authorised Yamaha dealer. It is not practicable to attempt to overhaul the crankshaft without the necessary press and trueing equipment. The Yamaha dealer will also be able to check and correct runout in a crankshaft that has become twisted or distorted for any reason. If the owner possesses a dial gauge and stand, runout may be checked by supporting the crankshaft on its centre bearings and then measuring deflection as the crank is rotated with the dial gauge needle resting on the end of each mainshaft and on both outer main bearings. No one reading should exceed 0.05 mm (0.0020 in). Correction of excessive runout requires a large degree of skill and experience if the problem is not to be made worse by the operator misunderstanding the cause of the problem. For this reason, professional help should be enlisted.

7   Failure of both the big-end bearings and the main bearings may not necessarily occur as the result of high mileage covered. If the machine is used only infrequently, it is possible that condensation within the engine may cause premature bearing failure. The condition of the flywheels is usually the best guide. When condensation troubles have occurred, the flywheels will rust and become discoloured. Note too that lack of care when disturbing the cylinder head or barrels can allow coolant to find its way into the crankcase. This will soon corrode and destroy the bearings and should be avoided for obvious reasons.

16 Oil seals: examination and renovation

1   The crankshaft oil seals form one of the most critical parts in any two-stroke engine because they perform the dual function of preventing oil from leaking along the crankshaft and preventing air from leaking into the crankcase when the incoming mixture is under crankcase vacuum during induction.

2   Oil-seal failure is difficult to define precisely, although in most cases the machine will become difficult to start, particularly when warm. The engine will also tend to run unevenly and there will be a marked fall-off in performance, especially in the higher gears. This is caused by the intake of air into the crankcases which dilutes the mixture whilst it is in the

15.3 Examine main and big-end bearings for wear

crankcase, giving an exceptionally weak mixture for ignition.

3   It is unusual for the crankshaft seals to become damaged during normal service, but instances have occurred when particles of broken piston rings have fallen into the crankcases and lacerated the seals. A defect of this nature will immediately be obvious.

4   In view of the foregoing remarks it is recommended that the two crankshaft oil seals are renewed as a matter of course during engine overhaul.

17 Cylinder barrels: examination and renovation

1   The usual indication of badly worn cylinder barrels and pistons is piston slap, a metallic rattle that occurs when there is little or no load on the engine. If the top of the bore of the cylinder barrels is examined carefully, it will be found that there is a ridge on the thrust side, the depth of which will vary according to the amount of wear that has taken place. This marks the limit of travel of the uppermost piston ring.

2   Measure the bore diameter just below the ridge, using an internal micrometer. Compare this reading with the diameter at the bottom of the cylinder bore, which has not been subjected to wear. If the difference in readings exceeds 0.05 mm (0.002 in) the cylinder should be rebored and fitted with an oversize piston and rings.

3   Bore ovality should also be checked, the maximum allowable being 0.1 mm (0.0039 in). Given that the bores are within the above limits and that the pistons are in serviceable condition (see Section 18) the parts may be re-used. Ovality may be corrected to some extent by honing, provided that this does not cause the maximum piston to bore clearance to be exceeded. A Yamaha dealer or a reputable engineering company will be able to assist with honing work should this prove necessary.

4   If scoring of the cylinder walls is evident it will normally prove necessary to have it re-bored to the next oversize, though light scratching may sometimes be removed by careful honing or by judicious use of abrasive paper. If the latter approach is adopted be careful to avoid removing more than the absolute minimum of material. The paper should be applied with a rotary motion **never** up and down the bore, which would cause more problems than it solves. One of the proprietary 'glaze busting' attachments for use in electric drills can be used to good effect for this operation. Even where the bore is in good condition, the glaze busting operation should be undertaken prior to re-assembly. The light scratch marks around the bore surface assist in bedding in the rings and help initial lubrication by holding a certain amount of oil.

5 If reboring is necessary, obtain the pistons first, then have the boring done to suit the new pistons. Most Yamaha dealers have an arrangement with a local engineering company and will be able to get the reboring work carried out promptly.

6 Carefully remove any accumulated carbon deposits from the cylinder bore and ports, taking care not to damage the bore surface. It is recommended that the ports are cleaned completely but carefully, taking great care to avoid burring the edges of the ports where they enter the bore. To prevent the rings from becoming chipped or broken dress any burrs with fine emery paper.

7 It is inadvisable to attempt modification of the port sizes or profiles to obtain more power from the engine. Such modifications are feasible but should only be considered for racing purposes. Generally speaking, the changed characteristics of the engine would make it unwieldy for road use, and it should be noted that the machine's warranty would be invalidated.

8 Check the water passages for rust and scale. These may have built up, especially where the correct coolant has not been used. If necessary, scrape the passages clean using wire or an old screwdriver, taking care to flush out any debris. Bear in mind that any residual debris may clog the radiator or pump if it is not removed.

18.4 Clean ring grooves. Note ring locating pegs (arrowed)

18 Pistons and piston rings: examination and renovation

1 If a rebore if necessary, the existing pistons and piston rings can be disregarded because they will have to be replaced with their new oversize equivalents as a matter of course.

2 Remove all traces of carbon from the piston crowns, using a blunt-ended scraper to avoid scratching the surface. Finish off by polishing the crowns with metal polish, so that carbon will not adhere so readily in the future. Never use emery cloth on the soft aluminium.

3 Piston wear usually occurs at the skirt or lower end of the piston and takes the form of vertical streaks or score marks on the thrust face. There may also be some variation in the thickness of the skirt, in an extreme case.

4 The piston ring grooves may have become enlarged in use, allowing the rings to have greater side float. If the clearances exceed those given, the rings, and possibly the pistons, must be renewed.

Piston to ring clearances

	RD250LC	RD350LC
Top ring	0.02–0.06 mm	0.02–0.06 mm
	(0.0008–0.0024 in)	(0.0008–0.0024 in)
Second ring	0.02–0.06 mm	0.03–0.07 mm
	(0.0008–0.0024 in)	(0.0012–0.0028 in)

5 Piston ring wear is measured by removing the rings from the piston and inserting them in the cylinder, using the crown of a piston to locate them about 20 mm from the bottom of the bore. Make sure they rest squarely in the bore. Measure the end gap with a feeler gauge; if the gap exceeds that given below, the rings must be replaced.

Piston ring end gap (installed)

	RD250LC	RD350LC
Top ring	0.30–0.45 mm	0.30–0.45 mm
	(0.0118–0.0177 in)	(0.0118–0.0177 in)
Second ring	0.30–0.45 mm	0.30–0.50 mm
	(0.0118–0.0177 in)	(0.0118–0.0197 in)

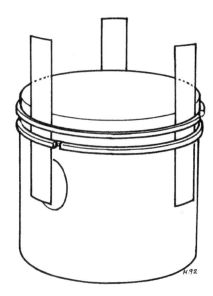

Fig. 1.6 Method of removing and replacing piston rings

19 Cylinder head: examination and renovation

1 Remove all traces of carbon from the cylinder head, using a blunt-ended scraper. Finish by polishing with metal polish, to give a smooth, shiny surface. This will aid gas flow and will also prevent carbon from adhering so firmly in the future.

2 Check the condition of the threads in the sparking plug holes. If the threads are worn or stretched as the result of overtightening the plugs, they can be reclaimed by a 'Helicoil' thread insert. Most dealers have the means of providing this cheap but effective repair.

3 Inspect the water passages cast into the cylinder head, and where necessary remove any accumulated corrosion or scale. As mentioned previously, this can result from failure to use the recommended coolant mixture. Be sure to remove any debris from the passages by flushing them through with clean water.

4 Lay the cylinder head on a sheet of plate glass to check for distortion. Aluminium alloy cylinder heads will distort very easily, especially if the cylinder head bolts are tightened down unevenly. If the amount of distortion is only slight, it is

permissible to run the head down until it is flat once again by wrapping a sheet of very fine emery cloth around the plate glass sheet and rubbing with a rotary motion.

5   If the cylinder head is distorted badly, it is advisable to fit a new replacement. Although the head joint can be restored by skimming, this will raise the compression ratio of the engine and may adversely affect performance.

20 Gearbox components: examination and renovation

1   Give the gearbox components a close visual inspection for signs of wear or damage such as broken or chipped teeth, worn dogs, damaged or worn splines and bent selectors. Replace any parts found unserviceable because they cannot be reclaimed in a satisfactory manner.

2   The gearbox shafts are unlikely to sustain damage unless the lubricating oil has been run low or the engine has seized and placed an unusually high loading on the gearbox. Check the surfaces of the shaft, especially where a pinion turns on it, and renew the shaft if it is scored or has picked up. The shafts can be checked for trueness by setting them up in V-blocks and measuring any bending with a dial gauge.

3   Examine the gear selector claw assembly noting that worn or rounded ends on the claw can lead to imprecise gear selection. The springs in the selector mechanism and the detent or stopper arm should be unbroken and not distorted or bent in any way.

4   The gearbox bearings must be free from play and show no signs of roughness when they are rotated. Each shaft has a ball journal bearing at one end and a caged needle roller bearing at the other.

5   It is advisable to renew the gearbox oil seals irrespective of their condition. Should a re-used oil seal fail at a later date, a considerable amount of dismantling is necessary to gain access and renew it.

6   Check the gear selector rods for straightness by rolling them on a sheet of plate glass. A bent rod will cause difficulty in selecting gears and will make the gear change action particularly heavy.

7   The selector forks should be examined closely, to ensure that they are not bent or badly worn. Wear is unlikely to occur unless the gearbox has been run for a period with a particularly low oil content.

8   The tracks in the gear selector drum, with which the selector forks engage, should not show any undue signs of wear unless neglect has led to under lubrication of the gearbox.

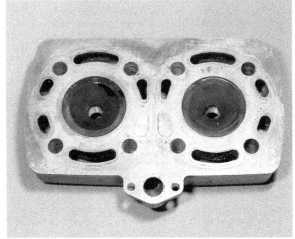

19.1 Clean combustion chambers and check head for warpage

20.8a Camplate is retained by a single cross-head screw

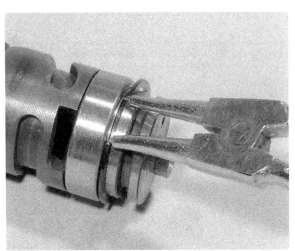

20.8b Remove large circlip from selector drum end ...

20.8c ... and slide the caged needle-roller bearing off

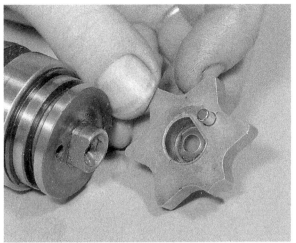

20.8d Note flat and locating pin during assembly

20.8e Neutral switch contact is secured by plate and screw

21 Kickstart mechanism: examination and renovation

1   The kickstart mechanism is a robust assembly and should not normally require attention. Apart from obvious defects such as a broken return spring, the friction clip is the only component likely to cause problems if it becomes worn or weakened. The clip is intended to apply a known amount of drag on the kickstart pinion, causing the latter to run up its quick thread and into engagement when the kickstart lever is operated.

2   The clip can be checked using a spring balance. Hook one end of the balance onto the looped end of the friction clip. Pull on the free end of the balance and note the reading at the point where pressure overcomes the clip's resistance. This should normally be 1.0 kg (2.2 lb). If the reading is higher or lower than this and the mechanism has been malfunctioning, renew the clip as a precaution. Do not attempt to adjust a worn clip by bending it.

3   Examine the kickstart pinion for wear or damage, remembering to check it in conjunction with the output shaft-mounted idler pinion. In view of the fact that these components are not subject to continuous use a significant amount of wear or damage is unlikely to be found.

21.1a Tension of kickstart friction clip should be checked

21.1b Plastic sleeve holds spring concentric to shaft

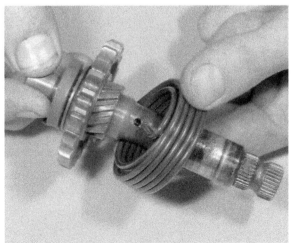

21.1c Inner tang of spring pushes into shaft drilling

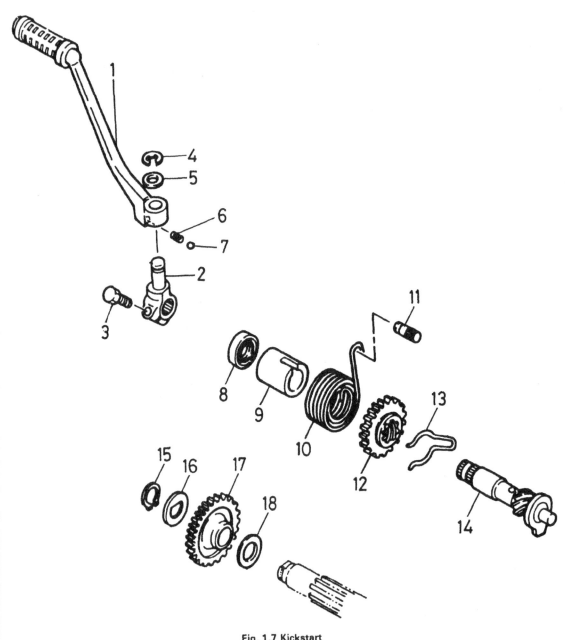

Fig. 1.7 Kickstart

1	Kickstart lever	7	Ball	13	Friction clip
2	Kickstarter lever knuckle	8	Oil seal	14	Kickstart shaft
3	Pinch bolt	9	Plastic spring guide	15	Circlip
4	Circlip	10	Return spring	16	Special washer
5	Washer	11	Anchor pin	17	Idler pinion
6	Spring	12	Kickstart pinion	18	Washer

22 Primary drive: examination and renovation

1   The primary drive consists of a crankshaft pinion which engages a large gear mounted on the inner face of the clutch drum. Both components are relatively lightly loaded and will not normally wear until very high mileages have been covered.
2   If wear or damage is discovered it will be necessary to renew the component concerned. In the case of the large driven gear it will be necessary to purchase a complete clutch drum because the two items form an integral unit and cannot be obtained separately.
3   When obtaining new primary drive parts note that the two components are matched to give a prescribed amount of backlash. To this end, ensure that the match marks marked on the inner face of each are similar to avoid excessive or insufficient clearance.

23 Clutch assembly: examination and renovation

1   After an extended period of service, the friction plates will have become worn sufficiently to warrant renewal, to avoid subsequent problems with clutch slip. The lining thickness is measured across the friction plate using a vernier caliper. When new, each plate measures 3.0 mm (0.118 in). If any plate is worn to 2.7 mm (0.106 in) or less the friction plates must be renewed.
2   The plain plates should be free from any signs of blueing, which would indicate that the clutch had overheated in the past. Check each plate for distortion by laying it on a flat surface, such as a sheet of plate glass or similar, and measuring any detectable gap using feeler gauges. The plates must be less than 0.05 mm (0.002 in) out of true.
3   The clutch springs may, after a considerable mileage, require renewal, and their free length should be checked as a precautionary measure. When new, each spring measures 34.9 mm (1.374 in) and the set should be renewed if they have compressed to 33.9 mm (1.335 in) or less.
4   Check the condition of the slots in the outer surface of the clutch centre and the inner surfaces of the outer drum. In an extreme case, clutch chatter may have caused the tongues of the inserted plates to make indentations in the slots of the outer drum, or the tongues of the plain plates to indent the slots of the clutch centre. These indentations will trap the clutch plates as they are freed and impair clutch action. If the damage is only slight the indentations can be removed by careful work with a file and the burrs removed from the tongues of the clutch plates in similar fashion. More extensive damage will necessitate renewal of the parts concerned.
5   The clutch release mechanism attached to the inside of the left-hand crankcase cover does not normally require attention, provided it is greased at regular intervals. It is held to the cover by two cross-head screws and operates on the worm and quick start thread principle. A light return spring ensures that the pressure is taken from the end of the clutch push rod when the handlebar lever is released and the clutch fully engaged.
6   Movement from the release mechanism is conveyed by a long push rod to the clutch pressure plate. Between the long push rod and the short mushroom-headed pushrod there is a single steel ball which allows for the rotation of the clutch. Check that these components are unworn and ensure that they are greased during reassembly. On very rare occasions where lubrication has been overlooked enough friction will have built up to break through the hardening on the pushrod ends. If this happens, the pushrod will tend to wear and may require frequent adjustment. The only satisfactory cure is to renew the affected parts. Finally, check that the long pushrod is straight by rolling it on a sheet of glass. If bent, renewal will be required.

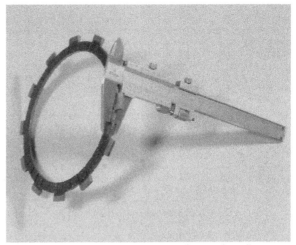

23.1 Measure friction plate thickness using vernier caliper

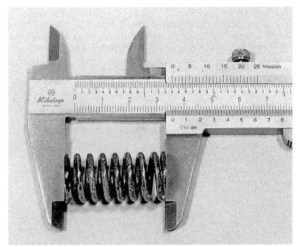

23.3 Check free length of clutch springs as shown

24 Engine reassembly: general

1   Before reassembly of the engine/gear unit is commenced, the various component parts should be cleaned thoroughly and placed on a sheet of clean paper, close to the working area.
2   Make sure all traces of old gaskets have been removed and that the mating surfaces are clean and undamaged. One of the best ways to remove old gasket cement is to apply a rag soaked in methylated spirit. This acts as a solvent and will ensure that the cement is removed without resort to scraping and the consequent risk of damage. If the gasket cement proves particularly stubborn it may be necessary to resort to using an aluminium or brass scraper. Do not use a screwdriver or a steel scraper because this will almost invariably damage the gasket face. One safe method is to use a brass wire brush such as those sold for cleaning suede shoes. This will usually prove very effective and will not damage the alloy.
3   Gather together all the necessary tools and have available an oil can filled with clean engine oil. Make sure all new gaskets and oil seals are to hand, also all replacement parts required. Nothing is more frustrating than having to stop in the middle of a reassembly sequence because a vital gasket or replacement has been overlooked.
4   Make sure that the reassembly area is clean and that there is adequate working space. Refer to the torque and clearance settings wherever they are given. Many of the smaller bolts are

easily sheared if over-tightened. Always use the correct sized screwdriver bit for the cross-head screws and never an ordinary screwdriver. If the existing screws show evidence of maltreatment in the past it is advisable to renew them as a complete set. It is strongly recommended that a set of Allen screws are used instead of the original cross-head screws. Allen screw sets can be obtained through most good accessory retailers and are an inexpensive but thoroughly practical improvement to most Japanese machines.

25 Engine reassembly: refitting the tachometer drive

1   Slide the tachometer drive shaft part way into the upper half of the crankcase and fit the innermost of the drive pinion retaining circlips. Fit the drive pinion over its splines and secure it with the remaining circlip. The shaft can now be pushed fully home, having lubricated it and the gear with clean engine oil.

2   On the outside of the casing, fit the retaining plate and screws to hold the shaft in position. Note that Loctite or a similar thread locking compound should be used on the two screws. Slide the drive gear locating pin through the shaft end, and place the gear over the end, securing it with its plain washer and circlip.

3   Assemble the tachometer driven gear and holder, having fitted a new O-ring to the latter where necessary. Slide the assembly into place and secure the single retaining bolt.

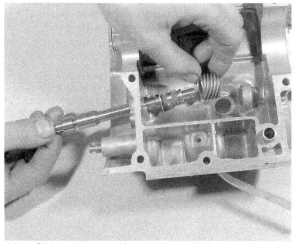

25.1a Slide tachometer drive shaft home, fitting inner circlip and drive gear

25.1b Gear is secured on shaft as shown

25.2a Retainer plate should engage in shaft groove

25.2b Push locating pin through hole in shaft ...

25.2c ... and place pinion as shown

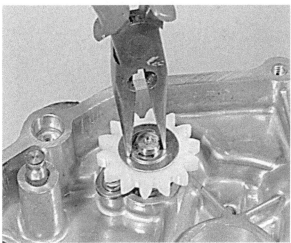

25.2d Pinion is secured by washer and retaining clip

25.3a Fit tachometer drive body, using new O-ring ...

25.3b ... and then tighten its single retaining bolt

their inner end so that when they are pushed fully home the circlips serve to locate the shaft ends.

5   Fit the selector drum retainer, noting that it also retains the front selector fork shaft. It is advisable to use Loctite on the two securing screws. Next, fit the selector mechanism centralising spring anchor pin, noting that the tab washer beneath it has an extended section which retains the rear selector fork shaft. The selector drum detent arm (stopper arm) can be fitted next and its pivot bolt tightened firmly. The selector shaft and claw assembly should be left off for the time being.

26 Engine reassembly: refitting the selector mechanism

1   Lubricate and fit the large needle roller bearing to the right-hand end of the selector drum and retain it with its circlip. Assemble the cam plate on the end of the drum, noting the small pin and the flat which locates it. Position the special cam retaining washer and secure the screw.

2   Moving to the left-hand end of the drum, fit the neutral switch plate assembly and retain it with its single screw. Make sure that the spring and contact are properly located.

3   Lubricate the plain (left-hand) end of the selector drum and slide it into position in the lower casing half. Do not fit the retainer plate at this stage. The selector forks and shafts should be fitted next. Note that two of the forks are identical, one being fitted to the front shaft and the other to the right-hand side of the rear shaft. The remaining fork is fitted to the left-hand end of the rear shaft.

4   Slide each shaft part way into the casing and fit the appropriate forks over it. The location pins should be arranged so that they engage with the selector drum tracks. Once the shafts are through the forks fit the circlips to the grooves on

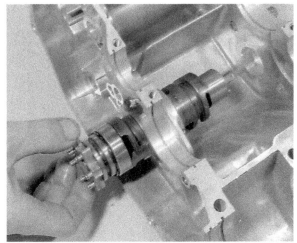

26.3a Lubricate and fit the gear selector drum

26.3b Fit rear selector forks and shaft, ...

26.3c ... securing shaft with clip as shown

26.3d Single front fork and shaft ...

26.3e ... are fitted in a similar fashion

26.5a Note that retainer plate locates shaft (arrowed)

26.5b Fit the eccentric pin and retainer plate

26.5c Assemble stopper arm and tighten pivot bolt

1 Selector drum
2 Front selector fork
3 Rear selector fork
4 Rear selector fork
5 Location pin – 3 off
6 Selector fork shaft
7 Circlip – 2 off
8 Selector fork shaft plug
   – 2 off
9 Selector fork shaft
10 Neutral switch plate
11 Spring
12 Neutral contact
13 Screw
14 Selector drum retainer
15 Screw – 2 off
16 Bearing
17 Circlip
18 Cam plate
19 Cam plate locating pin
20 Special washer
21 Screw
22 Spring
23 Selector drum stopper arm
24 Bolt

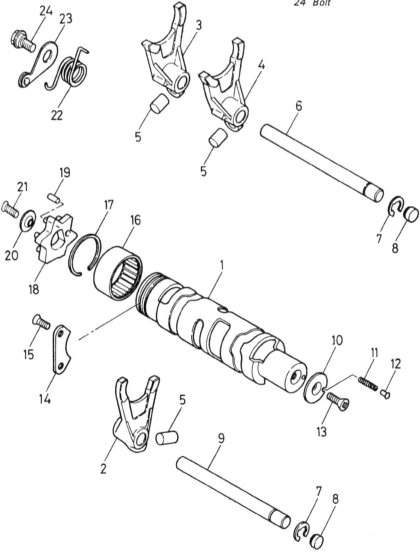

Fig. 1.8 Gearchange drum

27 Engine reassembly: refitting the gearbox components

1   Position the half rings which locate the two output shaft bearings in their grooves in the crankcase. A similar groove is provided in the recess for the input shaft's right-hand bearing, and this locates the bearing by means of the large circlip which is fitted to its outer race.

2   Lower the output shaft assembly into the crankcase lower half, ensuring that the selector fork fingers engage in the groove. The input shaft is now positioned in a similar manner. Check that both shafts seat securely.

3   Before proceeding further it is advisable to check gearbox operation. This can be done by turning the selector drum by means of the cam. To facilitate gear engagement, rotate the input shaft to and fro as each gear is selected. Neutral can be identified by noting that the detent arm drops into the shallowest of the cam depressions. When in neutral it should be possible to hold the output shaft stationary whilst the input shaft is turned. From neutral, select each gear in turn, ensuring that all six are available and that they each engage correctly.

28 Engine reassembly: refitting the crankshaft

1   The crankshaft should always be refitted using **new** oil seals. These are vital to the efficient running of all two-stroke engines. If a worn seal is reused, crankcase compression will be lost and performance will suffer. Before commencing reassembly, pack the gap between the seal lips with grease.

2   Position the locating half-ring in its groove in the right-hand main bearing boss. The half ring serves to locate the right-hand bearing, and thus the crankshaft. The remaining bearings are pegged to prevent rotation of the outer races, the pegs sitting in recesses to the front of each one. Lubricate each main bearing with new engine oil.

3   Lower the crankshaft into position, ensuring that the half ring and the three pegs are located correctly. Slide the seal into position over the crankshaft end. The seal should be positioned so that the outer face is flush with the crankcase boss, leaving a small gap between it and the main bearing. When correctly positioned the small bead around the outer face of the seal will locate in the corresponding groove in the casing recess.

4   Fit the right-hand seal in a similar manner, noting that it has a castellated spacing lip on its inner face. This should butt against the outer race of the main bearing, forming an additional method of location. When both seals are in place make sure that the crankshaft assembly is firmly seated along its length.

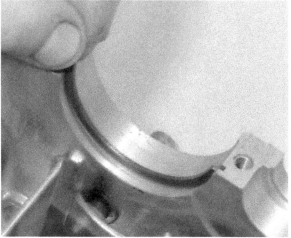
27.1 Fit half-ring into casing groove

27.2a Lower output shaft into position ...

27.2b ... then fit input shaft

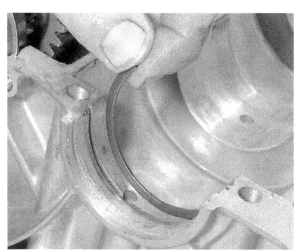
28.2a Half ring is fitted into crankcase groove ...

28.2b ... to locate right-hand main bearing

28.3a Centre bearings are located by small pins ...

28.3b ... as is the left-hand main bearing

28.4 RH oil seal must be fitted with castellations inwards

29 Engine reassembly: joining the crankcase halves

1   Make sure that the crankcase halves are clean and completely free of grease. To this end it is sound practice to give the jointing faces a final wipe with a clean rag moistened with methylated spirit or clean petrol. Allow the solvent to evaporate completely, then apply a thin film of jointing compound to the gasket face of one half. One of the RTV (room temperature vulcanising) silicone compounds, often sold as 'Instant Gasket' is recommended. Allow the compound to cure for a few minutes and in the meantime fit the two locating dowels to their recesses in the lower casing half.
2   The crankcase upper half can now be lowered into position, noting that the connecting rods must be fed through the crankcase apertures as the two halves meet. As the joint is closed, check that everything locates correctly, then tap the upper casing down with the palm of one hand to ensure that it locates firmly.
3   There are a total of 16 crankcase securing bolts, each of

which is numbered in the correct sequence for tightening. The numbers are cast into the crankcase next to the appropriate hole. When fitting the bolts it should be noted that bolts No 9, 14 and 15 have cable or wiring clips attached to them as shown in the tightening sequence diagram (Fig. 1.10).
4   Fit the upper crankcase bolts first (Nos 9 to 16) and tighten them just enough to secure the crankcase. Turn the unit over on the workbench and install the lower crankcase bolts (Nos 1 to 8). The bolts should now be tightened in two stages and in the sequence shown below:

a)   Bolts 9 to 16 to 0.5 kgf m (3.62 lbf ft)
b)   Bolts 1 to 8 to 1.0 kgf m (7.23 lbf ft)
c)   Bolts 1 to 8 to 2.5 kgf m (18.08 lbf ft)
d)   Bolts 9 to 16 to 1.0 kgf m (7.23 lbf ft)

5   The crankcases are now secured and before moving on, check that the crankshaft and the gearbox shafts rotate smoothly with no tight spots. If necessary, separate the crankcase halves and rectify any alignment problem before proceeding further.

29.1 Apply jointing compound and fit locating dowels

29.2 Upper crankcase half can be lowered into place

29.4a Fit crankcase upper bolts, noting wiring clips

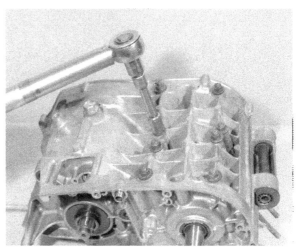
29.4b Lower crankcase is secured by eight bolts

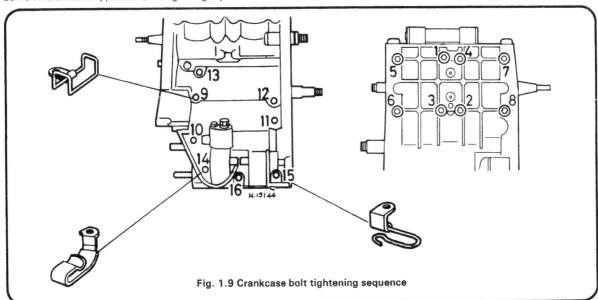
Fig. 1.9 Crankcase bolt tightening sequence

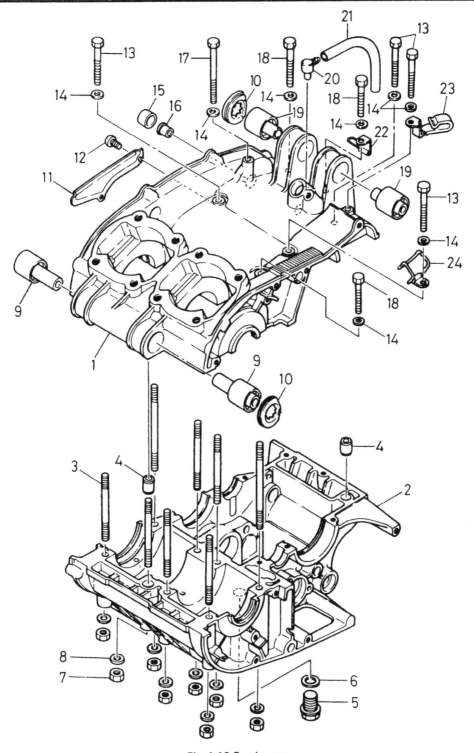

Fig. 1.10 Crankcases

1 Upper crankcase half	9 Front mounting bush – 2 off	17 Bolt
2 Lower crankcase half	10 Damping spacer – 2 off	18 Bolt – 3 off
3 Stud – 8 off	11 Side cover	19 Rear mounting bush – 2 off
4 Hollow dowel – 2 off	12 Screw – 2 off	20 Pipe cap
5 Drain plug	13 Bolt – 4 off	21 Pipe
6 Sealing washer	14 Washer – 8 off	22 Clamp
7 Nut – 8 off	15 Collar	23 Cable clip
8 Washer – 8 off	16 Bush	24 Cable clip

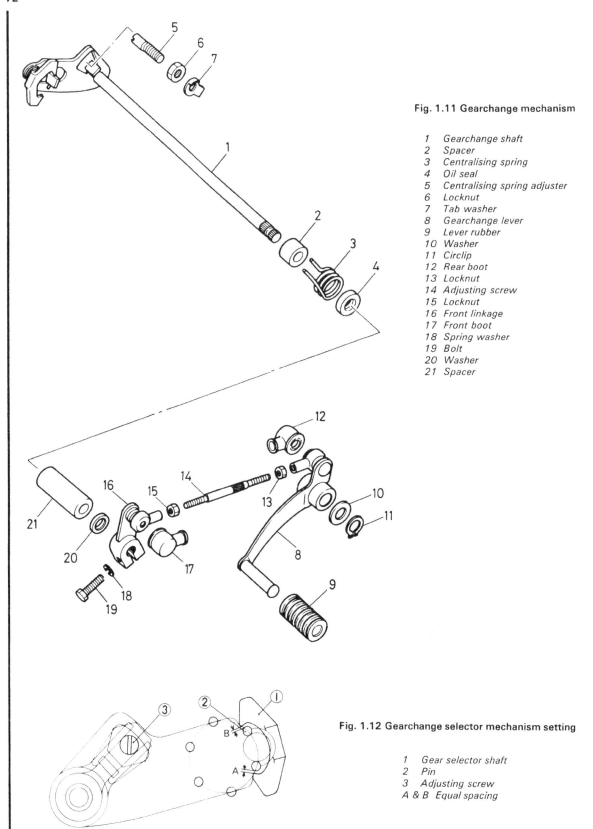

Fig. 1.11 Gearchange mechanism

1 Gearchange shaft
2 Spacer
3 Centralising spring
4 Oil seal
5 Centralising spring adjuster
6 Locknut
7 Tab washer
8 Gearchange lever
9 Lever rubber
10 Washer
11 Circlip
12 Rear boot
13 Locknut
14 Adjusting screw
15 Locknut
16 Front linkage
17 Front boot
18 Spring washer
19 Bolt
20 Washer
21 Spacer

Fig. 1.12 Gearchange selector mechanism setting

1 Gear selector shaft
2 Pin
3 Adjusting screw
A & B Equal spacing

30 Engine reassembly: fitting and adjusting the gear selector shaft

1   It is advisable to fit a new oil seal to the left-hand end of the gear selector shaft bore, irrespective of its condition. Lever the old seal out with a screwdriver, then tap the new seal into position using a suitably sized socket as a drift. Do not risk damaging the seal lip by hitting the seal directly. Lubricate the seal with a smear of grease before the shaft is installed. The oil seal in the engine left-hand cover should likewise be attended to if it is damaged.

2   Wrap some PVC tape around the splines on the selector shaft to protect the oil seal lip. Slide the shaft into its bore and check that the centralising spring ends engage on the eccentric adjusting screw. When at rest, the claw ends of the selector mechanism should be equidistant from the two adjacent pins (see Fig. 1.12). Check that this setting is correct in each gear and if necessary slacken the locknut on the centralising spring adjuster and adjust it to obtain the correct clearance.

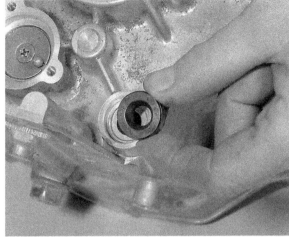

30.1a A new oil seal should be fitted to gearchange shaft bore

30.1b A new seal can also be fitted in outer cover

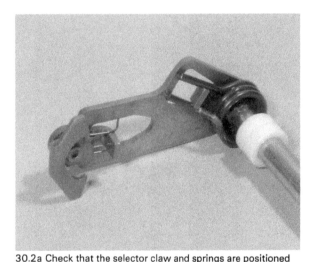

30.2a Check that the selector claw and springs are positioned as shown ...

31 Engine reassembly: fitting the kickstart mechanism idler pinion and crankcase fittings

1   Check that the kickstart pinion friction clip is in position, then slide the pinion over the shaft. Fit the kickstart return spring over the shaft and engage its inner tang in the shaft cross drilling. Once the spring is located slide the plastic spring guide into position to retain it. The assembly can now be fitted into the casing bore. Grasp the free end of the return spring and hook it over the anchor pin which protrudes from the upper casing half.

2   If it is not already in position the kickstart idler pinion should be fitted next. It is supported on the protruding end of the gearbox output shaft and is preceded by a plain washer. A special washer with an internal flat is fitted next and is secured by a circlip.

3   If it was removed during crankcase overhaul, refit the deflector plate above the input shaft bearing. It is retained by two cross-head screws, the threads of which should be coated with Loctite. The bearing retainer is fitted in a similar manner, noting that it bridges the crankcase halves. Where the cooling system stub was removed this should be refitted using a new O-ring and ensuring that its wire circlip is seated correctly.

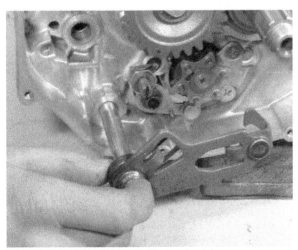

30.2b ... then slide selector mechanism into place

31.1a Note locating recess for friction clip tang (arrowed)

31.1b Hook return spring over anchor pin as shown

31.3 Fit deflector plate and input shaft bearing retainer

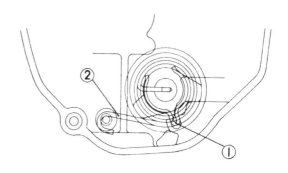

Fig. 1.13 Kickstart return spring position

1   *Friction clip*   2   *Return spring*

---

32 Engine reassembly: refitting the clutch, primary drive and pump drive pinion

---

1   Rotate the crankshaft until the keyway is uppermost, then fit the large Woodruff key. The crankshaft primary drive gear can now be slid into place, noting that its shouldered face should be completely smooth and well lubricated where it enters the oil seal. Take care not to force the seal lip inwards when fitting the pinion.

2   Slide the smaller pump drive pinion over the crankshaft end, then fit the Belville washer and securing nut. It is worth noting that although the pump pinion is relatively lightly loaded it is not keyed to the crankshaft and thus relies on the securing nut being tightened properly. If the nut becomes loose in service, the oil and water pumps would stop, followed swiftly by the engine which, unlubricated and uncooled, would soon seize. Lock the crankshaft as described during dismantling and tighten the securing nut to 6.5 kgf m (47 lbf ft).

3   Slide the large thrust washer over the end of the gearbox input shaft, followed by the clutch bush. The clutch drum can be fitted next, noting that it should engage with the primary drive and kickstart idler pinions. Fit the second thrust washer, followed by the clutch centre, tab washer and clutch centre nut. Lock the clutch, using the same method that was employed during dismantling, and tighten the nut to 6.5 kgf m (47 lbf ft).

4   The clutch plain and friction plates and the rubber damper rings should be coated with engine oil prior to installation. It will be noted that each of the plain plates has a part of its outer edge machined off. This effectively makes the plate become slightly out of balance. This causes each plate to be thrown outwards under centrifugal force and thus prevents clutch noise. To prevent the whole clutch from getting out of balance it is necessary to arrange the plates so that the machined areas are spaced evenly around its circumference. This can be achieved by arranging each cutaway area to be approximately 60° from the previous one.

5   Start by sliding a damper ring over the clutch centre, taking care not to twist it during fitting. A friction plate is fitted next, followed by a plain plate, this process being repeated until all the clutch plates are in position.

6   Slide the long pushrod through the hollow input shaft, noting that the end with the reduced diameter should be fitted first. The single steel ball can be pushed into the shaft bore now, followed by the mushroom-headed pushrod. Offer up the clutch pressure plate, aligning one of its three arrow marks with the corresponding mark on the clutch centre. Fit the clutch springs and secure the assembly by tightening the clutch bolts evenly and firmly in a diagonal sequence.

7   Check that the mating surfaces of the crankcase and outer cover are clean and dry and fit the locating dowel to its recess. Place a new gasket in position. Lubricate the primary drive and pump drive pinions, then offer up the cover, ensuring that the oil and water pump drives align correctly. Note that a smear of grease around the cooling system stub will facilitate installation. Fit the securing screws remembering to include the two cable clips. Though of largely academic interest, the correct torque setting for the securing screws is 1.0 kgf m (7.2 lbf ft).

8   Carry out clutch adjustment as described in Routine maintenance, unless of course the engine is removed from the frame, in which case carry out adjustment after it has been refitted and the clutch cable reconnected.

32.1a Place Woodruff key in crankshaft keyway ...

32.1b ... then fit primary gear over crankshaft end

32.2a Place pump drive pinion over crankshaft end ...

32.2b ... followed by Belville washer

32.2c Lock crankshaft and tighten securing nut

32.3a Fit large thrust washer (arrowed) and clutch bush

32.3b Clutch drum can now be slid into place

32.3c Fit plain thrust washer ...

32.3d ... followed by clutch centre

32.3e Fit tab washer and clutch centre nut

32.3f Lock crankshaft and tighten to prescribed torque

32.3g Do not omit to lock nut by bending up tab

32.5a Commence clutch assembly with damper ring

32.5b Place friction plate into clutch drum, followed by ...

32.5c ... the first plain plate, noting cutaway portion

32.5d Cutaway of second plain plate is positioned 60° away from first

32.6a Slide long pushrod into mainshaft bore ...

32.6b ... then fit the single steel ball ...

32.6c ... and the mushroom-headed pushrod

32.6d Clutch pressure plate can now be fitted ...

32.6e ... and the springs, washers and bolts secured

32.7a Fit dowels and a new clutch cover gasket

32.7b Cover can now be offered up and secured

33 Engine reassembly: refitting the alternator, neutral switch and left-hand outer cover

1   Push the alternator stator wiring through its hole in the crankcase and locate the wiring grommet to retain it. The stator can now be offered up. Align the temporary timing marks which were made during removal and fit the three retaining bolts finger tight. The ignition timing will now be approximately correct, but **must** be checked and where necessary reset once the engine has been started.

2   Fit the Woodruff key to the crankshaft keyway, ensuring that it seats correctly. Position the alternator rotor, then fit the plain washer, spring locking washer and securing nut. Lock the crankshaft and tighten the nut to 8.0 kgf m (58 lbf ft).

3   Check that the gearbox is in neutral, and where necessary temporarily refit the gear change pedal assembly and select neutral. Offer up the plastic neutral switch cover, ensuring that its contact aligns with the neutral contact on the end of the selector drum. Fit and tighten the three retaining screws. Feed the neutral switch lead behind the stator and through its slot in the casing wall, pushing the rubber guide block into place. Connect the lead to the neutral switch terminal.

4   Slide the plastic spacer sleeve over the protruding end of the selector shaft. It should be noted that although the outer cover can be fitted at this stage it is best left off as a reminder that the ignition timing must be checked.

34 Engine reassembly: refitting the pistons, cylinder barrels and cylinder head

1 Check that the crankcase mouth area is clean and free from grease deposits, then place the cylinder base gaskets over the holding studs. Lubricate the big-end and main bearings with two-stoke oil. Turn the crankshaft to TDC and pack clean rag around the connecting rods so that the crankcase mouths are covered. Lubricate the small-end bearings and slide them into position in the connecting rod eyes.

2   When fitting the pistons it is important to note that they must be fitted in the bore from which they were removed, unless the engine has been rebored in which case new pistons will be fitted. Note that each piston crown carries an arrow mark which should face forward. If the gudgeon pins are tight in the piston bosses it is a good idea to warm the pistons prior to fitting. This will cause the alloy piston to expand more than the steel pin and will make assembly much easier. Hot water at or near boiling point is the best way of heating them with no risk of distortion, but be wary of burns or scalding when using this method – use heavy gloves or some thick rag when handling the hot pistons.

3   Fit each piston in turn, locating the gudgeon pin with **new** circlips. It is false economy to risk reusing old circlips. They may appear to be in good order, and in practice may be quite satisfactory, but in view of their low cost do not run the risk of a weakened circlip breaking or working loose in service. Once the pistons are in place, lubricate the rings with clean two-stroke oil and check that the ring end gaps coincide with the locating pegs.

4   Each barrel has a tapered lead-in at its base to help in guiding the rings into the bore. As the barrel is pushed down over the piston use one hand to feed the rings into the bore. It is important to check that the barrels are exactly square to the crankcase and connecting rods, otherwise there is some risk of ring breakage where the ring ends pass close to the notch in the inlet port (see photograph). This is an important point and warrants the removal of the reed valve units so that a visual check can be made. Once the piston rings have entered the bores correctly the rag padding may be removed from the crankcase mouths and the barrel pushed firmly down onto the base gasket.

5   Check that the mating surfaces of the cylinder head and barrels are clean and dry, then place the cylinder head gasket in position. Do not use jointing compound on either surface. Offer up the cylinder head and fit the eight retaining bolts finger tight. The head bolts should be tightened in the sequence indicated by the numbers cast into the cylinder head. Initial tightening should be to about 1.0 kgf m (7.2 lbf ft). It will now be necessary to repeat the tightening operation, this time to the final value of 2.4 kgf m (17.4 lbf ft). Note that the cylinder head bolts must be re-tightened after the engine has been run and allowed to cool down. Refit the cast radiator hose union to the top of the cylinder head, using a new sealing gasket. Ensure that the three bolts are tightened evenly to avoid distortion.

6   Fit the temperature gauge sender housing to the cylinder head, using a new gasket. Fit the hose and adaptor between the cylinder head and the crankcase stub. The reed valve assemblies can now be refitted noting that the rubber adaptors must be renewed if they have cracked around the balance pipe stubs. Push the balance pipe into position.

33.1 Offer up the stator and fit the retaining bolts

33.2a Fit Woodruff key and position rotor ...

33.2b ... followed by plain washer, spring washer and nut

33.2c Lock crankshaft and tighten nut to prescribed torque

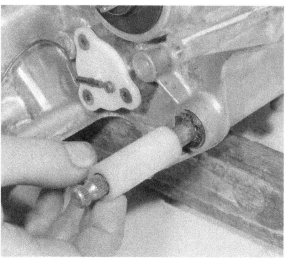

33.3 Connect neutral switch lead and fit gearchange sleeve

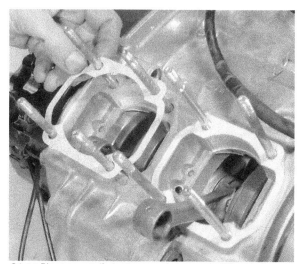

34.1a Place new cylinder base gaskets over holding studs

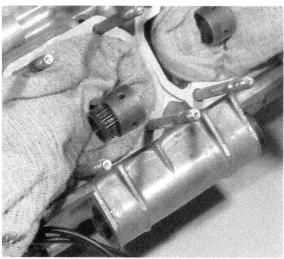

34.1b Pack crankcase mouths with rag and fit bearings

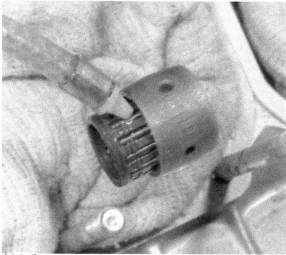

34.1c Small-end bearings should be lubricated as shown

34.2 Pistons are marked L and R. Arrow denotes front

34.4a Feed piston into cylinder bore ...

34.4b ... avoiding catching ring ends in port

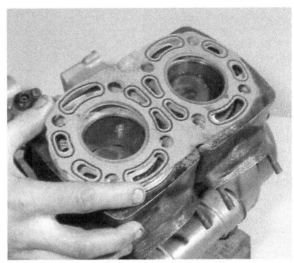

34.5a Place a new cylinder head gasket in position ...

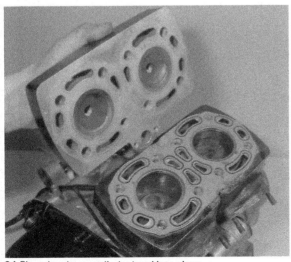

34.5b ... then lower cylinder head into place

34.5c Tighten sleeve bolts in the correct sequence

34.5d Fit the hose union, using a new gasket

35 Fitting the engine/gearbox unit in the frame

1   The engine/gearbox unit is refitted by reversing the removal sequence. As with removal, it is advantageous to have an assistant to help manoeuvre the unit into position, but the job is just about feasible unaided if this proves unavoidable. Take care not to damage the paintwork during installation. This can be guarded against by wrapping some card or stiff paper around the more vulnerable areas and taping this in place.
2   Lift the unit into the frame cradle from the right-hand side. It will sit in this position while the mounting brackets are sorted out and positioned. Before the engine is secured, make sure that the final drive chain is looped around the projecting cast boss to the rear of the output shaft. Failure to check this will cause problems later on since it is difficult to get the chain into position with the engine bolted into place.
3   Fit the front and rear mountings in position, fitting the small frame mounting bolts and the large through bolts finger tight. When all are in position, tighten the small bolts to 2.4 kgf m (17.4 lbf ft) and the large through bolts to 6.5 kgf m (47.0 lbf

ft). On late machines (1981 onwards) fit the engine steady bars between the frame brace tube and the underside of the crankcase.
4   Refit the tachometer drive cable to its adaptor at the rear of the crankcase. The knurled retaining ring should be tightened securely by hand. Fit the final drive chain around the gearbox sprocket and slide the assembly over the splined end of the output shaft. Place the tab washer against the sprocket, then fit the retaining nut noting that its recessed face should be against the tab washer. Lock the rear wheel by applying the brake, then tighten the nut to 6.5 kgf m (47.0 lbf ft). Bend the tab washer over one of the nut's flats and tap it down securely with a hammer and punch.
5   Connect the clutch cable to the left-hand outer cover, securing the nipple by bending over the small tang which prevents it from becoming displaced from the clutch release mechanism. Place some grease in the recess in the centre of the release mechanism, then fit the outer cover. Remove the circular inspection cover to reveal the clutch adjuster and carry out clutch adjustment as described in Routine Maintenance. Refit the cover, then assemble the remote gearchange linkage as shown in the accompanying photograph. Do not forget to check that the lever is set at the correct angle for convenient operation, and lubricate the pedal pivot with grease during installation.
6   Refit the throttle valves and caps to their respective carburettors. The valves and bodies are handed, but it is possible to get them interchanged and reversed. Guard against this by ensuring that the synchronisation dot on each valve will coincide with the small window in the carburettor body.
7   Manoeuvre the instruments into position between the airbox and inlet adaptors, and secure the retaining clips. Reconnect the oil pump cable and fit the pipe from the oil tank and the two small delivery pipes. The pump cover should be left off until it has been bled and adjusted.
8   Refit the exhaust system using new exhaust port sealing rings of the appropriate type. See Chapter 3 for further information regarding exhaust system modifications. Fit the radiator to the frame, but leave the hose to the water pump disconnected and the radiator guard off. Fit the spark plug caps and refit the temperature gauge sender lead.
9   Route the alternator output leads back along to the connector blocks below the seat. Assemble the halves of the connectors, using the colour-coded wires for identification. Fit and reconnect the battery, observing the correct polarity.

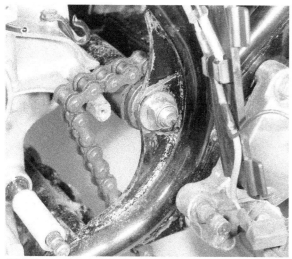

35.2 Ensure that chain is fitted around casing lug

35.3a Assemble the front engine plates and bolts ...

35.3b ... followed by rear mountings

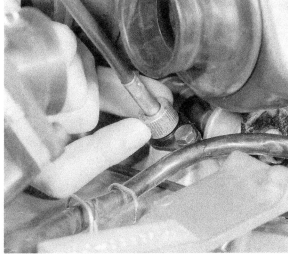

35.4a Connect tachometer cable to drive

35.4b Assemble sprocket, noting recessed face of nut

35.4c Lock rear wheel and secure sprocket nut ...

35.4d ... then bend over locking tab to secure it

35.5a Bend over security tab A. Grease recess B

35.5b Fit cover and adjust the clutch

35.5c Fit gearchange pedal, retaining it as shown ...

35.5d ... and secure front of linkage to shaft

35.5e Mechanism can be adjusted to alter pedal height

35.7a Reconnect pump hoses and cable ...

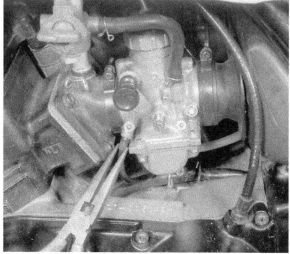
35.7b ... and fit delivery pipes to carburettors

35.8a Fit exhaust port sealing rings ...

35.8b ... and retain spacers with grease

35.8c Use new exhaust flange gaskets

36 Engine reassembly: final connections and adjustments

1   Check throttle cable free play and where necessary adjust to give 3 – 7 mm (0.12 – 0.28 in) movement measured at the outer edge of the twistgrip flange, making any adjustment with the in-line adjuster immediately below the throttle twistgrip.

2   Check throttle synchronisation by observing the alignment marks through the inspection windows on the right-hand side of each instrument. Using the adjusters on the carburettor tops, set both throttle valves so that the marks are central in their windows. Open and close the throttle a few times, then re-check.

3   Once synchronisation has been set, check the oil pump cable adjustment as described in Chapter 3, noting that the correct alignment mark is dependent on the model and year of manufacture. The oil pump should now be bled by removing the small bleed screw and allowing the air to be expelled by oil flowing from the tank. When the oil is free of air bubbles, fit and tighten the bleed screw. The oil delivery pipes should be bled once the engine is running as described in Section 37 of this Chapter.

4   Reconnect the hose to the water pump stub on the outer cover and fill the cooling system using a mixture of 50% distilled water and 50% Glycol antifreeze. Do not use ordinary tap water because the impurities contained in it will promote corrosion and furring-up of the system. Fit the radiator cap and the guard, but do not fit the two right-hand retaining screws. This will allow the guard to be displaced to permit topping up of the cooling system.

5   Remove the transmission oil filler plug and add 1700 cc (2.99 Imp pint) of SAE 10W//30 SE motor oil. The level of the oil should be re-checked after the engine has been run.

6   The ignition timing should be set using a dial gauge as described in Chapter 4. If this equipment is not available it will be necessary to assume that the reference marks made prior to alternator removal are reasonably accurate. It is essential that the timing is checked as soon as the machine is running, and this must be arranged with a local dealer if it cannot be done at home. There is a significant risk of engine damage if this operation is ignored.

7   Fit and secure the left-hand outer cover, and fit the kickstart lever if this is not already in position. Complete reassembly by fitting the fuel tank and pipe and the dual seat. Check around the machine to ensure that all remaining cables and connectors are in place. Where appropriate, readjust rear chain play and secure the wheel spindle nut. Make a final check of the electrical system by turning the ignition switch on and testing the operation of the various electrical components.

37 Starting and running the rebuilt engine

1   Initial starting may prove a little difficult and it is possible that the oil used during reassembly may cause fouling of the spark plugs. Use the normal cold starting procedure and be prepared for flooding during the first few attempts. If necessary remove and dry the plugs and start again. When the initial start-up is made, run the engine slowly for the first few minutes, especially if the engine has been rebored or a new crankshaft fitted. Check that all the controls function correctly and that there are no oil leaks, before taking the machine on the road. The exhausts will emit a high proportion of white smoke during the first few miles, as the excess oil used whilst the engine was reassembled is burnt away. The volume of smoke should gradually diminish until only the customary light blue haze is observed during normal running. It is wise to carry a spare pair of spark plugs during the first run, since the existing plugs may oil up due to the temporary excess of oil.

2   As soon as the engine is running evenly bleed the oil delivery lines by pulling on the pump cable so that the pump stroke is at maximum and the engine is held at a fast idle speed.

3   Remember that a good seal between the pistons and the cylinder barrels is essential for the correct functioning of the

engine. A rebored two-stroke engine will require more careful running-in, over a longer period, than its four-stroke counter-part. There is far greater risk of engine seizure during the first hundred miles if the engine is permitted to work hard.

4   Do not tamper with the exhaust system or run the engine without baffles fitted to the silencer. Unwarranted changes in the exhaust system will have a very marked effect on engine performance invariably for the worse. The same advice applies to dispensing with the air cleaner or the air cleaner element.

5   Do not on any account add oil to the petrol under the mistaken belief that a little extra oil will improve the engine lubrication. Apart from creating excess smoke, the addition of oil will make the mixture much weaker, with the consequent risk of overheating and engine seizure. The oil pump alone should provide full engine lubrication.

6   Before taking the machine on the road, check the dynamic ignition timing (see Chapter 4). It will also be necessary to allow the engine to cool down after its initial start-up. The cylinder head bolt torque and all oil and water levels should be re-checked. Replace all filler caps and covers and secure the right-hand side of the radiator guard. Remember to check the operation of all controls and electrical accessories before taking the machine on the road.

Chapter 2 Cooling system

Contents

General description 1
Draining the cooling system 2
Cooling system: flushing 3
Cooling system: filling 4
Radiator and radiator cap: removal, cleaning, examination
and refitting 5
Hoses and connections: examination and renovation 6
Water pump: removal and overhaul 7
Water temperature gauge and sender: testing 8

Specifications

Cooling system capacity .... 1.8 litre (3.8 Imp pint)

Coolant mixture .... 50% Distilled water, 50% Inhibited ethylene glycol anti-freeze

Radiator core size
    Width .... 272.5 mm (10.73 in)
    Height .... 180.0 mm (7.08 in)
    Thickness .... 32.0 mm (1.26 in)

Water pump
    Type .... Centrifugal impeller
    Drive .... Gear from pump idler pinion

Radiator pressure cap .... Opening pressure: 0.9 kg/cm^2 (12.8 psi)

1 General description

The Yamaha LC models are provided with a liquid cooling system which utilises a water/antifreeze coolant to carry away excess energy produced in the form of heat. The cylinders are surrounded by a water jacket from which the heated coolant is circulated by thermo-syphonic action in conjunction with a water pump fitted in the engine right-hand cover and driven via a pinion and shaft from a crankshaft mounted pinion. The hot coolant passes upwards through flexible pipes to the top of the radiator which is mounted on the frame downtubes to take advantage of maximum air flow. The coolant then passes downwards, through the radiator core, where it is cooled by the passing air, and then to the water pump and engine where the cycle is repeated.

The complete system is sealed and pressurised; the pressure being controlled by a valve contained in the spring loaded radiator cap. By pressurising the coolant the boiling point is raised, preventing premature boiling in adverse conditions. The overflow pipe from the radiator is connected to an expansion tank into which excess coolant is discharged by pressure. The expelled coolant automatically returns to the radiator, to provide the correct level when the engine cools again.

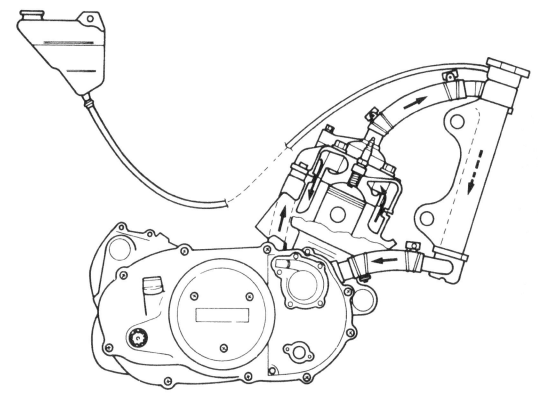

Fig. 2.1 Cooling system

2   Draining the cooling system

1   It will be necessary to drain the cooling system on infrequent occasions, either to change the coolant at two yearly intervals or to permit engine overhaul or removal. The operation is best undertaken with a cold engine to remove the risk of scalding from hot coolant escaping under pressure.

2   Place the machine on its centre stand and gather together a drain tray or bowl of about 2.0 litres (4.0 pint) capacity, and something to guide the coolant from the cylinder barrel drain plugs into the bowl. A small chute made from thick card will suffice for this purpose, but do not be tempted to allow the coolant to drain over the engine casings – the anti-freeze content may discolour the painted surfaces.

3   Remove the two screws which retain the right-hand side of the radiator guard. This will allow the guard to be pulled away from the radiator just enough for the radiator cap to be released. Take great care when removing the cap from the radiator if the engine has been run recently, because there will be some residual pressure in the system. If the engine is hot, steam and boiling water may be ejected and can cause scalding. As a precaution, place some rag over the cap and remove it slowly to allow pressure to escape.

4   Slacken and remove each of the cylinder barrel drain plugs in turn, using the chute to guide the coolant clear of the crankcase outer covers and into the bowl. If the system is to be drained fully pull off the pipe from the expansion tank and allow this to drain. Drain any residual coolant by detaching the hose at the front of the right-hand outer cover stub.

5   If the system is being drained as a precursor to engine overhaul little else need be done at this stage. If the coolant is

reasonably new it can be re-used if it is kept clean and uncontaminated. If, however, the system is to be refilled with new coolant it is advisable to give it a thorough flushing with tap water, if possible using a hose which can be left running for a while. If the machine has done a fairly high mileage it may be advisable to carry out a more thorough flushing process as described below.

2.3 Radiator shroud is retained by four screws

2.4a Remove cylinder drain plugs to drain coolant

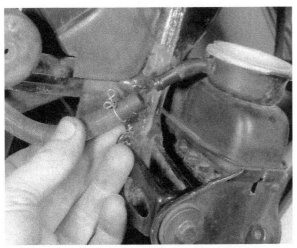

2.4b Expansion tank pipe is push fit on radiator stub

2.4c Release hose to drain residual coolant

3 Cooling system: flushing

1   After extended service the cooling system will slowly lose efficiency, due to the build up of scale, deposits from the water and other foreign matter which will adhere to the internal surfaces of the radiator and water channels. This will be particularly so if distilled water has not been used at all times. Removal of the deposits can be carried out easily, using a suitable flushing agent in the following manner.
2   After allowing the cooling system to drain, refit the drain plugs and refill the system with clean water and a quantity of flushing agent. Any proprietary flushing agent in either liquid or dry form may be used, providing that it is recommended for use with aluminium engines. NEVER use a compound suitable for iron engines as it will react violently with the aluminium alloy. The manufacturer of the flushing agent will give instructions as to the quantity to be used.
3   Run the engine for ten minutes at operating temperatures and drain the system. Repeat the procedure TWICE and then again using only clean cold water. Finally, refill the system as described in the following Section.

4 Cooling system: filling

1   Before filling the system, check that the sealing washers on the drain plugs are in good condition and renew if necessary. Fit and tighten the drain plugs and check and tighten all the hose clips.
2   Fill the system slowly to reduce the amount of air which will be trapped in the water jacket. When the cooling level is up to the lower edge of the radiator filler neck, run the engine for about 10 minutes at 900 rpm. Increase engine revolutions for the last 30 seconds to accelerate the rate at which any trapped air is expelled. Stop the engine and replenish the coolant level again to the bottom of the filler neck. Refill the expansion tank up to the 'Full' level mark. Refit the radiator cap, ensuring that it is turned clockwise as far as possible.
3   Ideally, distilled water should be used as a basis for the coolant. If this is not readily available, rain water, caught in a non-metallic receptacle, is an adequate substitute as it contains only limited amounts of mineral impurities. In emergencies only, tap water can be used, especially if it is known to be of the soft type. Using non-distilled water will inevitably lead to early 'furring-up' of the system and the need for more frequent flushing. The correct water/antifreeze mixture is 50/50; do not allow the antifreeze level to fall below 40% as the anti-corrosion properties of the coolant will be reduced to an unacceptable level. Antifreeze of the ethylene glycol based type should always be used. Never use alcohol based antifreeze in the engine.

5 Radiator and radiator cap: removal, cleaning, examination and refitting

1   Drain the radiator as described in Section 2 of this Chapter.
2   Disconnect the top hose at the cylinder head union by loosening the screw clip. Release the lower hose at the right-hand engine casing stub. Remove the two remaining radiator grille screws and lift the guard clear.
3   The radiator is secured to the frame by four rubber-mounted bolts. When removing these, note the relative position of the rubber and metal washers as a guide during reassembly. When the bolts have been released the radiator can be lifted clear of the frame.
4   Remove any obstructions from the exterior of the radiator core, using an air line. The conglomeration of moths, flies and autumnal detritus usually collected in the radiator matrix severely reduces the cooling efficiency of the radiator.

Fig. 2.2 Radiator

1   Radiator
2   Radiator guard
3   Grommet – 4 off
4   Collar – 4 off
5   Spring washer – 4 off
6   Bolt – 4 off
7   Screw – 4 off
8   Radiator cap
9   Hose clamp – 2 off
10 Lower hose
11 O-ring
12 Circlip
13 Hose union
14 Hose clamp – 2 off
15 Lower hose
16 Engine casing stub
17 Gasket
18 Bolt – 2 off
19 Gasket
20 Cylinder head union
21 Bolt – 2 off
22 Bolt
23 Washer
24 Hose clamp – 2 off
25 Top hose

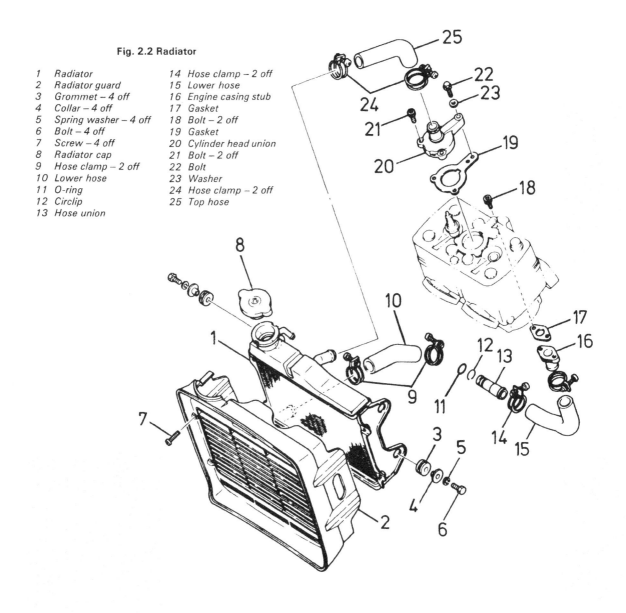

5   The interior of the radiator can most easily be cleaned while the radiator is in-situ on the motorcycle, using the flushing procedure described in Section 3 of this Chapter. Additional flushing can be carried out by placing the hose in the filler neck and allowing the water to flow through for about ten minutes. Under no circumstances should the hose be connected to the filler neck mechanically as any sudden blockage in the radiator outlet would subject the radiator to the full pressure of the mains supply (about 50 psi). The radiator should not be tested to greater than 1.0 kg/cm^2 (15 psi).

6   If care is exercised, bent fins can be straightened by placing the flat of a screwdriver either side of the fin in question and carefully bending it into its original shape. Badly damaged fins cannot be repaired. If bent or damaged fins obstruct the air flow more than 20%, a new radiator will have to be fitted.

7   Generally, if the radiator is found to be leaking, repair is impracticable and a new component must be fitted. Very small leaks may sometimes be stopped by the addition of a special sealing agent in the coolant. If an agent of this type is used, follow the manufacturer's instructions very carefully. Soldering, using soft solder may be efficacious for caulking large leaks but this is a specialised repair best left to experts.

8   Inspect the four radiator mounting rubbers for perishing or compaction. Renew the rubbers if there is any doubt as to their condition. The radiator may suffer from the effect of vibration if the isolating characteristics of the rubber are reduced.

10 If the radiator cap is suspect, have it tested by a Yamaha dealer. This job requires specialist equipment and cannot be done at home. The only alternative is to try a new cap.

5.2 Upper hose is retained by screw clip

5.3a Remove the radiator mounting bolts ...

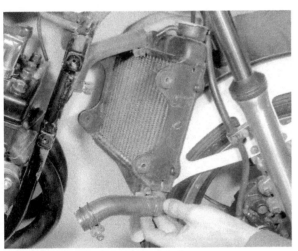

5.3b ... and lift the radiator clear of the frame

to be pinpointed. To this end it is best to entrust this work to an authorised Yamaha dealer who will have access to the necessary equipment.

3 In very rare cases the leak may be due to a broken head gasket, in which case the coolant may be drawn into the engine and expelled as vapour in the exhaust gases. If this proves to be the case it will be necessary to remove the cylinder head for investigation. If the rate of leakage has been significant it may prove necessary to remove the cylinder barrels and pistons so that the crankcase can be checked. Any coolant which finds its way that far into the engine can cause rapid corrosion of the main and big-end bearings and must be removed completely.

4 Another possible source of leakage is the stub between the crankcase and the right-hand outer cover. If its O-ring seal becomes damaged or broken it is possible that coolant might find its way into the transmission, and any sign of emulsified transmission oil or water droplets inside the cover should be investigated promptly before corrosion takes place.

6 Hoses and connections: examination and renovation

1 The radiator is connected to the engine unit by two hoses, there being an additional hose between the water pump in the right-hand outer cover and the cylinder head. The hoses should be inspected periodically and renewed if any sign of cracking or perishing is discovered. The most likely area for this is around the wire hose clips which secure each hose to its stub. Particular attention should be given if regular topping up has become necessary. The cooling system can be considered to be a semi-sealed arrangement, the only normal coolant loss being minute amounts through evaporation in the expansion tank. If significant quantities have vanished it must be leaking at some point and the source of the leak should be investigated promptly.

2 Serious leakage will be self-evident, though slight leakage can be more difficult to spot. It is likely that the leak will only be apparent when the engine is running and the system is under pressure, and even then the rate of escape may be such that the hot coolant evaporates as soon as it reaches the atmosphere. Such small leaks may require the use of a special device which will pressurise the system whilst cold and thus enable the leak

6.4 O-ring must be in good condition to avoid leaks

7 Water pump: removal and overhaul

1   The water pump will not normally require attention unless its bearing has become noisy if there is obvious leakage of coolant into the transmission oil. To gain access to the pump, drain the coolant and the transmission oil fully, then remove the pump cover followed by the right-hand outer cover itself. Carefully drain the residual coolant from the pump before dismantling commences.

2   The water pump is located immediately above the oil pump, the two sharing a common device pinion on the crankshaft. To dismantle the water pump it will first be necessary to remove its driven gear by displacing the circlip which secures it to the pump spindle. The locating pin should be removed by pushing it through and out of the spindle.

3   Unscrew the five cover screws and remove the cover and gasket. The impeller and spindle may now be displaced and removed. If the bearing or oil seal is worn or damaged they should be renewed as a set. The two components can be driven out from the oil seal side, having first heated the casing in an oven to 90° – 120°C (194° – 248°F). If using this method it is best to remove all seals, plastic parts and the oil pump first.

4   An alternative method is to pour boiling water (100°C) over the bearing boss area, but it is not advisable to use a blowlamp or other localised heat source in view of the risk of warpage.

Once heated, the bearing and seal can be driven out using a suitable round bar or an old socket.

5   The new bearing and oil seal should be greased prior to installation and tapped home using a large socket against the outer race of the bearing. Note that the seal is marked WATER SIDE on one face, and this should face the pump. The bearing serial number should face outward. Tap the bearing and seal home ensuring that they both seat squarely in the casing.

6   Clean the impeller and spindle, being particularly careful to ensure that any corrosion that may have formed around the seal area is removed and the spindle left completely smooth. If the spindle is badly pitted in this area it may be necessary to renew it to avoid rapid seal wear. The spindle should be greased prior to installation and care should be exercised during fitting to avoid damage to the seal face. Complete reassembly by reversing the dismantling sequence, using a new gasket on the pump cover joint. Where practicable, tighten the cover screws to 0.7 kgf m (5.1 lbf ft).

8 Water temperature gauge and sender: testing

1   Water temperature is monitored by an electrically operated gauge in the instrument panel controlled by a sender unit which screws into the cylinder head water jacket. A description and test procedure of these components will be found in Chapter 7.

7.2a Water pump is mounted above oil pump

7.2b Prise off clip and remove white plastic pinion ...

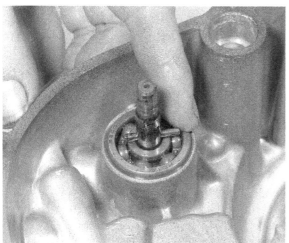

7.2c ... then displace drive pin from shaft end

7.3a Pump cover is secured by five screws

7.3b Pump shaft and impeller can be displaced

7.3c Drive out the old oil seal and bearing as shown

7.5a Use socket to drive in new bearing and seal

7.5b Ensure that seal locates properly in bore

7.6a Use new gasket at pump cover joint

7.6b Fit pin, pinion and retain with washer and clip

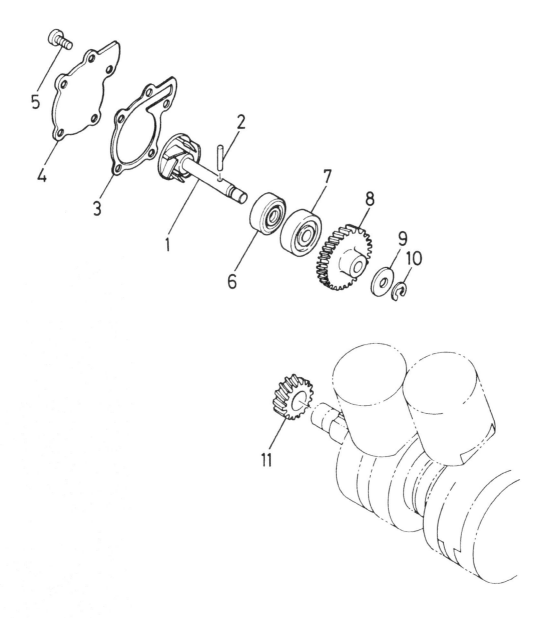

Fig. 2.3 Water pump

1	Drive shaft	6	Oil seal
2	Locating pin	7	Bearing
3	Cover gasket	8	Driven gear
4	Pump cover	9	Washer
5	Screw – 5 off	10	Circlip
		11	Drive gear

Chapter 3 Fuel system and lubrication

Contents

General description 1
Petrol tank : removal and replacement 2
Petrol tap : removal, dismantling and reassembly 3
Petrol feed pipes : examination 4
Carburettors : general description 5
Carburettors : removal and refitting 6
Carburettors : dismantling and reassembly 7
Carburettors : adjustment 8
Carburettor settings 9

Exhaust system : cleaning 10
Exhaust system : modifications 11
Air cleaner : removal and cleaning 12
The engine lubrication system 13
Removing and replacing the oil pump 14
Bleeding the oil pump 15
Checking the oil pump and throttle cable settings 16
Reed valve induction system : mode of operation 17
Reed valves : removal, examination and renovation 18

Specifications

Carburettors

	1980 models up to 4L1/4L0 100101*	
	RD250 LC	**RD350 LC**
Make	Mikuni	Mikuni
Type	VM26SS	VM26SS
ID mark	4L100	4L000
Main jet	190	160 *see note a*
Air jet	1.0	0.5
Jet needle	4N10	4H16
Clip position	4	2
Needle jet	O-6	O-6
Throttle valve cutaway	2.0	2.0
Pilot jet	20.0	27.5 *see note b*
Air screw, turns out	$1\frac{1}{8}$	$1\frac{1}{2}$ *see note c*
Starter jet	80	80
Fuel level	26 ± 1 mm	26 ± 1 mm
	(1.02 ± 0.04 in)	(1.02 ± 0.04 in)
Float height	21 ± 0.5 mm	21.0 ± 0.5 mm
	(0.83 ± 0.02 in)	(0.83 ± 0.02 in)
Idle speed	1200 ± 50 rpm	1200 ± 50 rpm

Notes:

a) *140 main jets optional – superseded.*

b) *Early models had 25 pilot jets with 27.5 optional. Models from engine No 4L0 – 004669 were fitted with 27.5 pilot jets in conjunction with 1V0 reed valves.*

c) *Air screw setting 1.0 turn out on early models*

* *A number of modifications were incorporated on 1980 RD350LC models to eliminate a mid-range misfire problem. These are described in the main text, but owners should always check which parts have been fitted to their particular machine rather than rely solely on these specifications. Refer to Section 5 for further details.*

Carburettors

	Engine Nos 4L1/4L0 100101 onwards	
	RD250 LC	RD350 LC
Make	Mikuni	Mikuni
Type	VM26SS	VM26SS
ID mark	Not available	Not available
Main jet	190	220
Air jet	1.0	0.8
Jet needle	4N10	5K1
Clip position	4	3
Needle jet	O-6	P-2 (345)
Throttle valve cutaway	2.0	2.0
Pilot jet	20.0	22.5
Air screw, turns out	$1\frac{1}{8}$	$1\frac{1}{2}$
Starter jet	80	80
Fuel level	Not available	Not available
Float height	21 mm	21 mm
Idle speed	1200 rpm	1200 rpm

Engine lubrication

Type	Pump fed total loss system (Yamaha Autolube)
Oil tank capacity	1.6 litre (2.8 Imp pint)
Pump minimum stroke setting:	
1980 models	0.20-0.25 mm (0.008-0.010 in)
Modified 1980 models	0.15 mm (0.006 in)
1981 models	0.10-0.15 mm (0.004-0.006 in)

1 General description

1   The fuel system comprises a petrol tank, from which petrol is fed by gravity to the float chambers of the twin Mikuni carburettors, via a three position petrol tap. The tap has 'Off', 'On' and 'Reserve' positions, the latter providing a warning that the petrol level is low in time for the owner to find a garage.

2   The carburettors are of conventional concentric design, the float chambers being integral with the lower part of the carburettor bodies. Cold starting is assisted by a separate starting circuit which supplies the correct fuel-rich mixture when the 'choke' control is operated. The cold start mechanism is fitted to the left-hand carburettor only.

3   Air entering the carburettors passes through a moulded plastic air cleaner casing, which contains an oil-impregnated foam air filter. This effectively removes any airborne dust, which would otherwise enter the engine and cause premature wear. The air cleaner also helps silence induction noise, a common problem inherent with two-stroke engines.

4   Engine lubrication is catered for by the Yamaha Autolube system. Oil from a separate tank is fed by an oil pump to small injection nozzles in the inlet tract. The pump is linked to the throttle twistgrip, and this controls the volume of oil fed to the engine.

5   The exhaust system is a two piece affair, each cylinder having its own expansion chamber with integral exhaust pipe. The system is finished in a heat-resistant matt black coating.

6   It should be noted that the carburettors, oil pump and the exhaust system have all been subject to modification since the LC range was first introduced. Where practicable, this has been mentioned in the text of this Chapter, but it must be stressed that an authorised Yamaha dealer must be consulted when replacement parts are required. This will ensure that the correct parts are supplied and any necessary modifications have been carried out.

2 Petrol tank : removal and replacement

1   It is unlikely that the petrol tank will need to be removed except on very infrequent occasions, because it does not restrict access to the engine unless a top overhaul is to be carried out whilst the engine is in the frame.

2   The petrol tank is secured at the rear by a single bolt, washer and rubber buffer that threads into a strut welded across the two top frame tubes. It is necessary first to remove the dual seat before access is available.

3   When the bolt and washer are withdrawn, the petrol tank can be lifted from the frame. The nose of the tank is a push fit over two small rubber buffers, attached to a peg that projects from each side of the frame, immediately to the rear of the steering head. A small rubber 'mat' cushions the rear of the tank and prevents contact with the two top frame tubes.

4   The petrol tank has a locking filler cap to prevent pilferage of the tank contents when the machine is left anattended.

3 Petrol tap : removal, dismantling and reassembly

1   The petrol tap is secured to the underside of the tank by two hexagon-headed screws. There is seldom need to disturb the main body of the petrol tap. In the event of a leak at the operating lever, the complete lever assembly can be dismantled (provided the petrol tank is drained first) with the main body undisturbed.

2   The tap lever assembly can be withdrawn for inspection, after releasing the single crosshead screw which locates it. If leakage has been evident, the most likely culprit will be the O-ring which seals the tap valve against the body. If this fails to effect a cure, it will be necessary to renew the complete tap assembly.

3   The tap is provided with a sediment bowl, in which any fine debris from the tank which has managed to get through the filter gauze, will be trapped, along with any water. The bowl should be periodically removed for cleaning. When refitting the sediment bowl, ensure that the O-ring is in good condition.

4   Before reassembling the petrol tap, check that all the parts are clean, especially the tube which forms the filter and main and reserve intakes.

5   Do not overtighten any of the petrol tap components during reassembly. The castings are in a zinc-based alloy, which will fracture easily if over-stressed. Most leakages occur as the result of defective seals.

3.1 Fuel tap is secured to underside of tank by two screws

4 Petrol feed pipes : examination

The petrol feed pipes, connecting the carburettors to the fuel tap, are made of thin walled plastic, and are retained by small wire clips. Check that the pipes have not split or become brittle, due to age and the effects of heat and fuel. Check also the various drain and breather pipes.

5 Carburettors : general description

All models to date have employed variations of the Mikuni VM26SS carburettors, with a number of jetting permutations as shown in the specifications section of this Chapter.

Early models of the RD350LC were prone to a mid-range misfire, and the initial modification to resolve this problem was to fit 140 main jets and a different reed valve assembly. This proved to be only partially sucessful, and a second modification was adopted in which a 3 mm air bleed was drilled through the carburettor body to intersect with the air jet passage. The original orifice was then blanked off with Araldite epoxy adhesive. Carburettors modified in this way were re-jetted as shown below:

Reed valve	4L0-13610-00 or 1V0-13610-00
Pilot jet	25 or 27.5
Needle clip position	3rd notch
Air screw	$\frac{3}{4}$-$1\frac{1}{4}$ turns out
Main jet	160 standard (170 optional for continous high speed)

It should be noted that machines with engine numbers from 4L0-000101 to 4L0-004668 were fitted with the 4L0 reed valve and 25 pilot jet, whilst engine number 4L0-004669 onward had the 1V0 reed valve and 27.5 pilot jet. *All* machines from engine number 4L0 100101 onwards (1981 models) had all necessary modifications incorporated during manufacture and thus required no attention from the dealer. The above modification should, if required, be carried out by an authorised Yamaha dealer and no attempt should be made to drill the carburettors at home.

6 Carburettors : removal and refitting

1 As a general rule, the carburettors should be left alone unless they are in obvious need of overhaul. Before a decision is made to remove and dismantle them, ensure that all other possible sources of trouble have been eliminated. This includes the more obvious candidates such as fouled spark plugs, a dirty air filter element or chocked exhaust system baffles. If a fault has been traced back to the carburettors, proceed as follows.
2 Make sure that the fuel tap is turned off, then prise off the petrol feed pipes at the carburettor stubs. The oil delivery pipes are removed in a similar manner, noting that the small tubular clips should be displaced first. The pipes can then be eased away from their stubs with the aid of an electrical screwdriver.
3 Slacken the screws of the clips which secure each carburettor to its inlet and airbox adaptors. Each carburettor can now be twisted free of the rubber adaptors and partially removed. This affords access to the threaded carburettor tops, which should be unscrewed to allow the throttle valve assemblies to be withdrawn. It is not normally necessary to remove these from the cables, and they can be left attached and taped clear of the engine. If removal is necessary, however, proceed as follows.
4 Holding the carburettor top, compress the throttle return spring against it and hold it in position against the cap. Invert the throttle valve and shake out the pressed steel spring seat. This component serves to prevent the cable from becoming detached when in position and once out of the way the cable can be pushed down and slid out of its locating groove. The various parts can now be removed and should be placed with the instrument to which they belong. Do not allow the parts to be interchanged between the two instruments.
5 The carburettors are refitted by reversing the removal sequence. Note that it is important that the instruments are mounted vertically to ensure that the fuel level in the float bowls is correct. A locating tab provides a good guide to alignment but it is worthwhile checking this for accuracy. Once refitted, check the carburettor adjustments and synchronisation as described later in this Chapter. Note too that the oil pump delivery pipes should be bled and the pump adjustments checked after overhaul.
6 **Note:** if the carburettors are to be set up from scratch it is important to check jet and float level settings prior to installation. To this end, refer to the next two Sections before the carburettors are refitted.

6.2 Prise off oil and fuel pipes (arrowed)

6.3a Slacken mounting clips and displace carburettors

6.3b Unscrew carburettor top and withdraw throttle valve

6.4 Compress spring and remove retainer to free cable

7 Carburettors : dismantling and reassembly

1   Invert each carburettor and remove the float chamber by withdrawing the four retaining screws. The float chamber bowls will lift away, exposing the float assembly, hinge and float needle. There is a gasket between the float chamber bowl and the carburettor body which need not be disturbed unless it is leaking.

2   With a pair of thin nose pliers, withdraw the pin that acts as the hinge for the twin floats. This will free the floats and the float needle. Check that none of the floats have punctured and that the float needle and seating are both clean and in good condition. If the needle has a ridge, it should be renewed in conjunction with its seating.

3   The two floats are made of plastic, connected by a brass bridge and pivot piece. If either float is leaking, it will produce the wrong petrol level in the float chamber, leading to flooding and an over-rich mixture. The floats cannot be repaired successfully, and renewal will be required.

4   The main jet is located in the centre of the circular mixing chamber housing. It is threaded into the base of the needle jet and can be unscrewed from the bottom of the carburettor. The needle jet lifts out from the top of the carburettor, after the main jet has been unscrewed. The pilot jet is located in a smaller projection, next to the main jet.

5   The float needle seating is also found in the underside of the carburettor, towards the bell mouth intake. It is secured by a small retainer plate and is sealed by an O-ring. If the float needle and the seating are worn, they should be replaced as a set, never separately. Wear usually takes the form of a ridge or groove, which may cause the needle to seat imperfectly.

6   The carburettor valves, return springs and needle assemblies together with the mixing chamber tops, are attached to the throttle cable. The throttle cable divides into two at a junction box located within the two top frame tubes. There is also a third cable, which is used to link the oil pump with the throttle.

7   After an extended period of service the throttle valves will wear and may produce a clicking sound within each carburettor body. Wear will be evident from inspection, usually at the base of the slide and in the locating groove. Worn slides should be replaced as soon as possible because they will give rise to air leaks which will upset the carburation.

8   The needles are suspended from the valves, where they are retained by a circlip. The needle is normally suspended from the groove specified at the front of this Chapter, but other grooves are provided as a means of adjustment so that the mixture strength can be either increased or decreased by raising or lowering the needle. Care is necessary when replacing the carburettor tops because the needles are easily bent if they do not locate with needle jets.

9   The manually operated choke is unlikely to require attention during the normal service life of the machine. When the plunger is depressed, fuel is drawn through a special starter jet in the left-hand carburettor by a partial vacuum that is created in the crankcase. Air from the float chamber passes through holes in the starter emulsion tube to aerate the fuel. The fuel then mixes with air drawn in via the starter air inlet to the plunger chamber. The resultant mixture, richened for a cold start, is drawn into the engine through the starter outlet, behind the throttle valve.

10   Before the carburettors are reassembled, using the reversed dismantling procedure, each should be cleaned out thoroughly, preferably by the use of compressed air. Avoid using a rag because there is always risk of fine particles of lint obstructing the internal air passages or the jet orifices.

11   Never use a piece of wire or sharp metal object to clear a blocked jet. It is only too easy to enlarge the jet under these circumstances and increase the rate of petrol consumption. Always use compressed air to clear a blockage; a tyre pump makes an admirable substitute when a compressed air line is not available.

12   Do not use excessive force when reassembling the carburettors because it is quite easy to shear the small jets or some of the smaller screws. Before attaching the air cleaner hoses, check that both throttle slides rise when the throttle is opened.

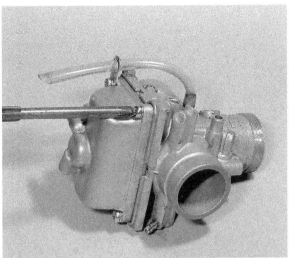

7.1a Float bowl is retained by four cross-head screws

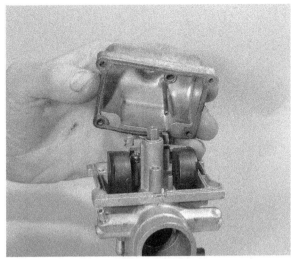

7.1b Remove float bowl taking care not to damage gasket

7.2a Displace float pivot pin and remove float

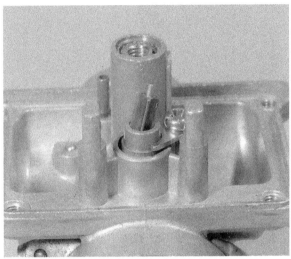

7.2b Valve can be tipped out of seating

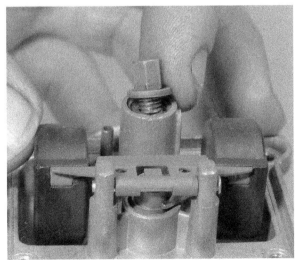

7.4a Main jet is screwed into base of needle jet

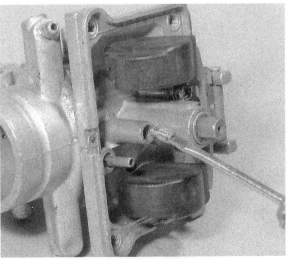

7.4b Pilot jet is fitted into adjacent drilling

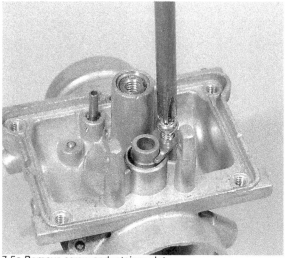
7.5a Remove screw and retainer plate ...

7.5b ... and withdraw float valve seat

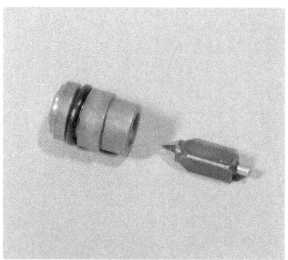
7.5c Check valve and seat condition as a pair

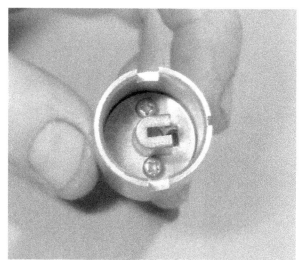
7.8a Remove screws to free throttle cable anchor

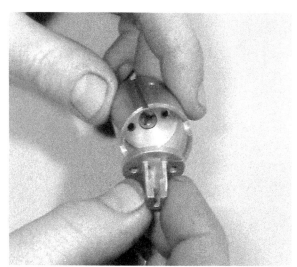
7.8b Lift anchor plate away and displace needle

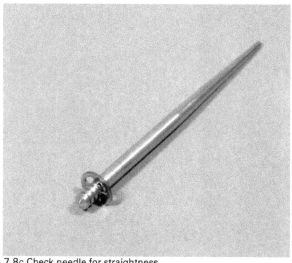
7.8c Check needle for straightness

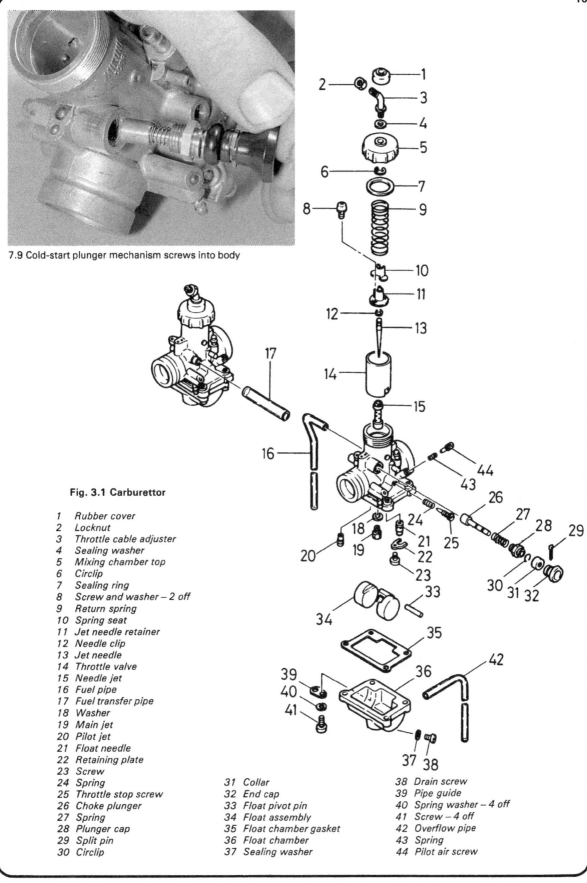

7.9 Cold-start plunger mechanism screws into body

Fig. 3.1 Carburettor

1   Rubber cover
2   Locknut
3   Throttle cable adjuster
4   Sealing washer
5   Mixing chamber top
6   Circlip
7   Sealing ring
8   Screw and washer – 2 off
9   Return spring
10 Spring seat
11 Jet needle retainer
12 Needle clip
13 Jet needle
14 Throttle valve
15 Needle jet
16 Fuel pipe
17 Fuel transfer pipe
18 Washer
19 Main jet
20 Pilot jet
21 Float needle
22 Retaining plate
23 Screw
24 Spring
25 Throttle stop screw
26 Choke plunger
27 Spring
28 Plunger cap
29 Split pin
30 Circlip

31 Collar
32 End cap
33 Float pivot pin
34 Float assembly
35 Float chamber gasket
36 Float chamber
37 Sealing washer

38 Drain screw
39 Pipe guide
40 Spring washer – 4 off
41 Screw – 4 off
42 Overflow pipe
43 Spring
44 Pilot air screw

8 Carburettors : adjustment

1   The first step in carburettor adjustment is to ensure that the jet sizes, needle position and float height are correct, which will require the removal and dismantling of the carburettors as described in Section 7. It should be noted that a number of modifications have been made to some 1980 models, and where there is doubt about the suitability of jet sizes or settings it is advisable to seek advice from an authorised Yamaha dealer.

2   Before any dismantling or adjustment is undertaken eliminate all other possible causes of running problems, checking in particular the spark plugs, ignition timing, air cleaner and the exhaust baffles. Checking and cleaning these items as appropriate will often resolve a mysterious flat spot or misfire.

3   If the carburettors have been removed for the purpose of checking jet sizes, the float level should be measured at the same time. It is unlikely that once this is set up correctly, there will be a significant amount of variation, unless the float needle or seat have worn. These should be checked and renewed as required. With the float bowl removed slowly rotate the carburettor until gravity acting on the floats moves the float until the valve is **just** closed, but not so far that the needle's spring-loaded pin is compressed. Measure the distance between the gasket face and the bottom of the float with an accurate ruler. The correct setting should be 21.0 ± 0.5 mm (0.827 ± 0.020 in). If adjustment is required it can be made by bending by a very small amount, the small tang to which the float needle is attached.

4   When the carburettors are being refitted, set the throttle stop screws as follows. Fit the throttle valve assemblies to their respective instruments and secure the carburettor tops before the bodies are fitted into their adaptors. Identify the throttle stop screws, which will be found projecting at right angles to the main body. These should be slackened off completely to allow the throttle valves to close fully. Check the throttle cable free play, measured at the throttle twistgrip flange and where necessary set this to 3-7 mm (0.118 - 0.276 in) cable free play should be set using the adjuster and locknut arrangement located immediately below the twistgrip housing.

5   The throttle stop screws should be screwed slowly inward until they **just** contact the underside of the valves. If this is done

carefully it should be possible to feel and see the point at which this occurs. It is worth spending a little time to ensure that the two screws are set accurately. An alternative method is to use a metal rod, the plain end of a drill bit being ideal. Set the throttle stop screws so that the rod is a light sliding fit beneath each throttle valve cut away. With either of the above methods, the object is to ensure that the two screws are set in similar positions. Once set it is important to ensure that in subsequent adjustments each screw is moved by exactly the same amount as the other. The two instruments can now be refitted to the machine.

6   Next, check that the two throttles are synchronised. Unless this is established it will prove impossible to persuade the engine to run evenly, and poor synchronisation will make mixture and throttle stop settings futile. Standing on the right-hand side of the machine open the throttle twist-grip fully and observe the two small windows in the side of each instrument body. A small alignment pip should be visible in each one, and it is important to arrange these so that they are in accurate synchronisation. Each valve can be adjusted via the independent adjuster in each mixing chamber top.

7   The remaining adjustments are made with the engine running and at normal operating temperature. To this end it may prove necessary to make some provisional idle speed adjustment, remembering to adjust each throttle stop screw by an equal amount to keep the carburettors in balance. Set the pilot air screws to the position shown in the specifications for the appropriate model. Note the pilot air screw on the right-hand carburettor is awkwardly located on its inner face and some degree of dexterity will be called for when adjustment is required. Taking each cylinder in turn, rotate the pilot air screw inwards and outwards from the datum setting until the position is found at which the engine runs fastest. Reduce the idle speed if necessary, and then adjust the second pilot air screw in a similar manner.

8   Check the engine idle speed and adjust both throttle screws equally to bring it to the specified speed. The carburettors should by now be fairly accurately set up and in many instances no further adjustment will be necessary. If, however, there appears to be room for improvement at idle speed, fine adjustment of the throttle stop screws should bring things into balance.

8.4 Throttle stop screw location

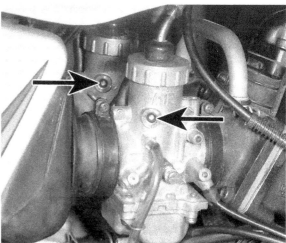

8.6a Windows in carburettor allow synchronisation mark to be seen

8.6b Synchronisation adjustment is made via cable adjusters

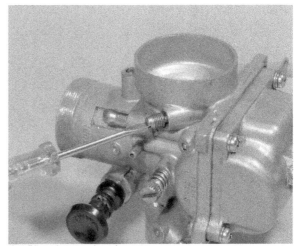

8.7 Pilot air screw location

9 Carburettor settings

1   Some of the carburettor settings, such as the sizes of the needle jets, main jets and needle position are predetermined by the manufacturer. Under normal circumstances it is unlikely that these settings will require modification, even though there is provision made. If a change appears necessary, it can often be traced to a developing engine fault.

2   As a rough guide, the slow running screw controls the engine speed up to $\frac{1}{8}$ throttle. The throttle valve cutaway controls the engine speed from $\frac{1}{8}$ to $\frac{1}{4}$ throttle and the position of the needle from $\frac{1}{4}$ to $\frac{3}{4}$ throttle. The main jet is responsible for the engine speed at the final $\frac{3}{4}$ to full throttle. It should be added that none of these demarkation lines is clearly defined; there is a certain amount of overlap between the carburettor components involved.

3   Always err on the side of a rich mixture because a weak mixture has a particularly adverse effect on the running of any two-stroke engine. A weak mixture will cause rapid overheating which may eventually promote engine seizure. Reference to Chapter 3 will show how the condition of the sparking plugs can be used as a reliable guide to carburettor mixture strength.

10 Exhaust system : cleaning

1   The exhaust system is often the most neglected part of any two-stroke despite the fact that it has a pronounced effect on performance. It is essential that the exhaust system is inspected and cleaned out at regular intervals because the exhaust gases from a two-stroke engine have a particularly oily nature which will encourage the build-up of sludge. This will cause back pressures and restrict the engine's ability to 'breathe'.

2   Cleaning is made easy by fitting the silencers with detachable baffles, held in position by a set screw that passes through each silencer end. If the screw is withdrawn, the baffles can be drawn out of position for cleaning.

3   A wash with a petrol/paraffin mix will remove most of the oil and carbon deposits, but if the build-up is severe it is permissible to heat the baffles with a blow lamp and burn off the carbon and old oil.

4   At less frequent intervals, such as when the engine requires decarbonising, it is advisable also to clean out the exhaust pipes. This will prevent the gradual build-up of an internal coating of carbon and oil, over an extended period.

5   Do not run the machine with the baffles detached or with a quite different type of silencer fitted. The standard production silencers have been designed to give the best possible performance whilst subduing the exhaust note. Although a modified exhaust system may give the illusion of greater speed as a result of the changed exhaust note, the chances are that performance will have suffered accordingly.

6   When replacing the exhaust system, use new sealing rings at the exhaust port joints and check that the baffle retaining screws are tightened fully in the silencer rods.

11 Exhaust system : modifications

The original LC models tended to suffer from fracturing around the exhaust port flanges due to the inability of the exhaust system to move adequately in response to the engine's movement in its mountings. In some cases, the pipes would not fracture but would move enough to cause leakage around the flange gaskets.

The initial modification was to remove the metal spacer from the center of the rear silencer mounting rubber and to reduce its length from 22mm to 20mm. This helped to make the assembly more rigid without losing the resilience of the rear mountings, but was only partially successful in overcoming the problem.

Subsequent modifications included the use of aluminium spacers in the exhaust ports and eventually a revised exhaust system for the earlier models. All 1981 machines feature the modified exhaust as standard. Should exhaust system sealing or fracture problems be encountered it is recommended that the advice of a Yamaha repair agent is sought. He will be able to advise on the latest available modified components and the best course of action for repair.

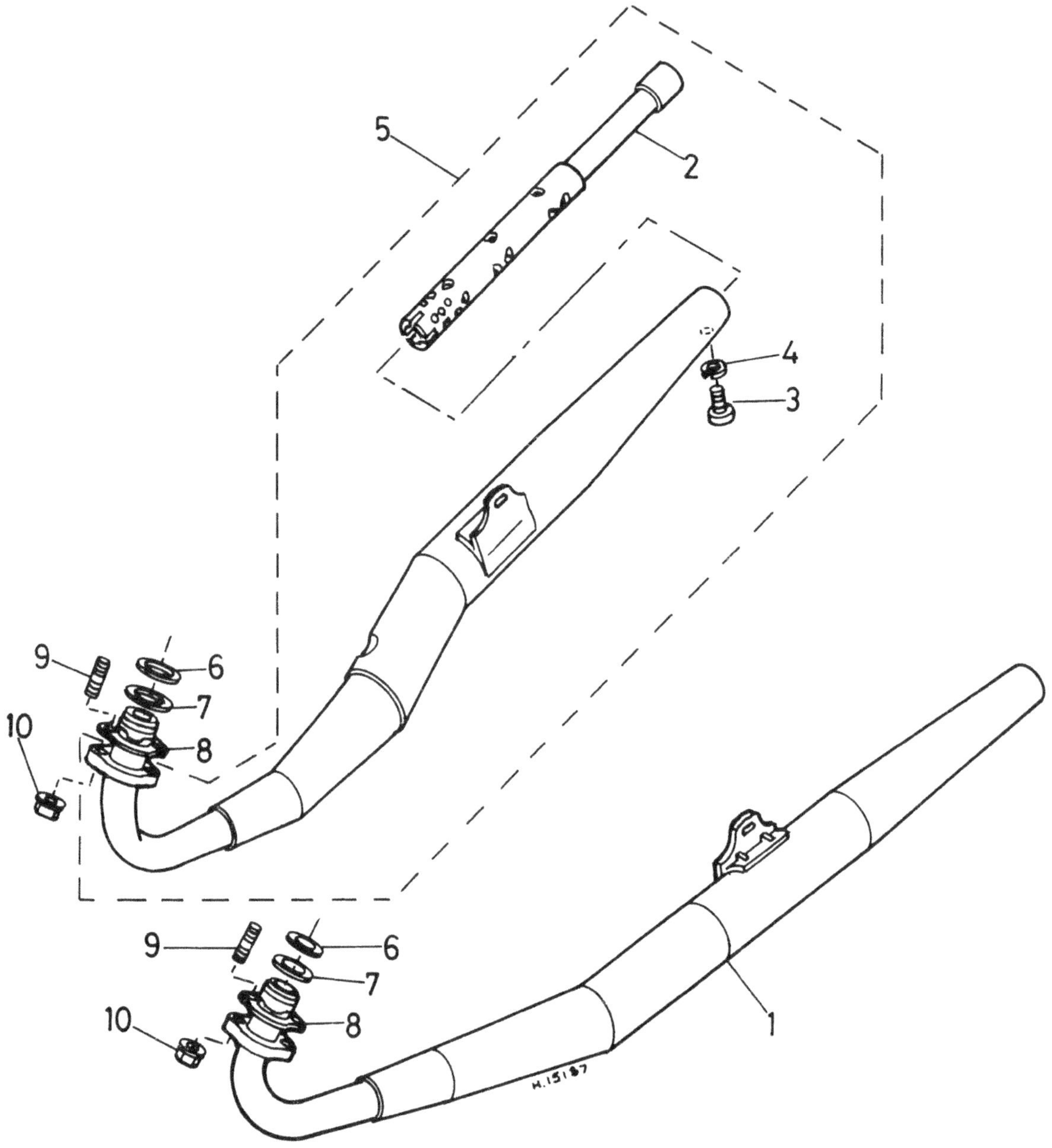

Fig. 3.2 Exhaust system

1	Left-hand exhaust pipe	6	Spacer -- 1980 models only
2	Baffle	7	Gasket
3	Screw	8	Flange gasket
4	Spring washer	9	Stud – 2 off
5	Right-hand exhaust pipe	10	Nut – 2 off

12 Air cleaner : removal and cleaning

1   The air cleaner casing is mounted on the frame beneath the fuel tank. It is connected to a second moulded plastic chamber immediately behind the carburettors. This functions as an intake silencer and conveys the cleaned air to the carburettor intakes via short rubber adaptors.

2   Access to the air cleaner element requires the removal of the fuel tank and seat. The seat can be lifted clear once its retaining latches have been released. Turn the fuel tap off and disconnect the fuel feed pipes. Remove the single fixing bolt at the rear of the tank and lift the tank upwards and rearwards to free it from its front mounting rubbers.

3   The air cleaner cover is retained by three screws. Remove the screws and lift the cover clear to expose the flat foam element. This can be removed in turn by pulling it out of the casing. Wash the element in clean petrol to remove the old oil and any dust which has been trapped by it. When it is clean, wrap it in some clean rag and gently squeeze out the remaining petrol. The element should now be left for a while to allow any residual petrol to evaporate. Soak the cleaned element in engine oil and then squeeze out any excess to leave the foam damp but not dripping. Refit the element, ensuring that the cover seat can now be refitted.

4   Note that a damaged element must be renewed immediately. Apart from the risk of damage from ingested dust, the holed filter will allow a much weaker mixture and may lead to overheating or seizure. It follows that the machine must never be used without the filter in position.

5   The rest of the air cleaner system requires little attention, other than checking that the connecting rubbers are undamaged. When checking these do not omit the adaptors which connect the carburettors to the inlet ports. These are prone to perishing and cracking around the balance pipe stubs and should be renewed if leakage is suspected.

12.3a Remove screws and lift air filter cover away

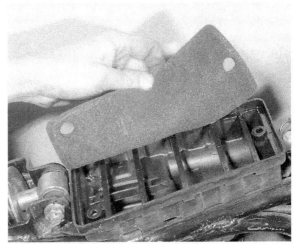
12.3b Filter element can be removed for cleaning

13 The engine lubrication system

1   In line with current two-stroke practice the Yamaha LC models utilise a pump-fed engine lubrication system and do not require the mixture of a measured quantity of oil to the petrol content of the fuel tank in order to utilise the so-called 'petroil' method. Oil of the correct viscosity is contained in a separate oil tank mounted on the left-hand side of the machine and is fed to a mechanical oil pump on the right-hand side of the engine which is driven from the crankshaft by reduction gear. The pump delivers oil at a predetermined rate, via two flexible plastic tubes, unions on the inlet side of the carburettor venturis. In consequence, the oil is carried into the engine by the incoming charge of petrol vapour, when the inlet port opens.

2   The oil pump is also interconnected by the twist grip throttle, so that when the throttle is opened, the oil pump setting is increased a similar amount. This technique ensures that the lubrication requirements of the engine are always directly related to the degree of throttle opening. This facility is arranged by means of a control cable attached to a lever on the ens of the pump; the cable is joined to the throttle cable junction box at the point where the cable splits into two for the operation of each carburettor.

3   On some of the earlier machines there were cases of excessive oil consumption. This was corrected by reducing the minimum pump stroke and setting the pump pulley to a new index mark as described in Section 16. On 1981 machines (Engine numbers 4L0/4L1 100101 onwards) the oil pump pulley is set to a different mark depending on the model and pump identification marks. Again, see Section 16 for details.

12.5a Balance pipe is fitted between inlet adaptors

12.5b Look for signs of cracking around balance pipe stubs

Fig. 3.3 Air cleaner

1 Screw – 3 off
2 Washer – 3 off
3 Top cover
4 Casing seal
5 Intake hose
6 Air cleaner element
7 Bolt
8 Washer
9 Air cleaner casing
10 Seal
11 Intake silencer
12 Nut – 2 off
13 Intake adaptor – 2 off
14 Retaining clamp – 2 off
15 Washer – 2 off
16 Bolt – 2 off

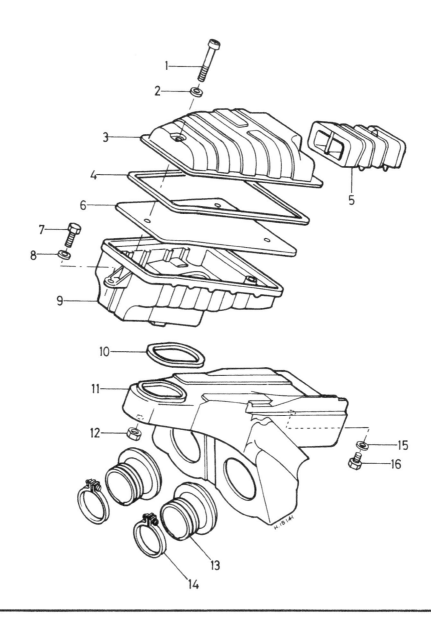

14 Removing and replacing the oil pump

1   It is rarely necessary to remove the oil pump unless specific attention to it is required. It should be noted that the pump should be considered a sealed unit – parts are not available and thus it is not practicable to repair it. The pump itself can be removed quite easily leaving the drive shaft and pinion in place in the right-hand outer casing. If these latter components require attention it will be necessry to drain the cooling system and transmission oil so that the right-hand outer casing can be removed. Refer to Chapter 1 for further details. The accompanying photographic sequence describes the procedure for removing the pump drive components.

2   To gain access to the oil pump, remove the screws which secure the pump cover to the right-hand outer casing. With these removed the pump will be clearly visible at the bottom of the pump recess. Do not disturb the water pump end cover which is immediately above the oil pump.

3   Displace the small spring steel clips which secure the oil delivery pipes to the pump outlets, then ease the pipes off the outlet stubs using a small screwdriver. The large feed pipe from the oil tank is removed in a similar fashion, but before removing it have some sort of plug handy to push into the end of the pipe. This will prevent the oil from the tank being lost. Pull on the pump cable inner to rotate the pump pulley. Holding the pulley in its fully open position release the cable and disengage it from the pulley recess.

4   The pump is secured to the cover by two screws which pass through its mounting flange. Once these have been removed the pump can be removed, noting that it may prove necessary to turn the pump slightly to free it from its drive shaft.

5   Further dismantling is not practicable, and it will be necessary to renew the pump if it is obviously damaged. Maintenance must be confined to keeping the pump clear of air, and correctly adjusted, as described in the following sections.

6   Refit the oil pump to the crankcase cover, using a new gasket at the oil pump/crankcase cover joint. Replace and tighten the two crosshead mounting screws. The remainder of the reassembly is accomplished by reversing the dismantling procedure, but do not replace the pump cover because the oil pump must be bled to ensure the oil lines are completely free from air bubbles. See the following Section.

14.1a Pump spindle pinion is secured by E-clip

14.1b Remove pump pinion ...

14.1c ... and displace drive pin

14.1d Spindle can now be removed from casing

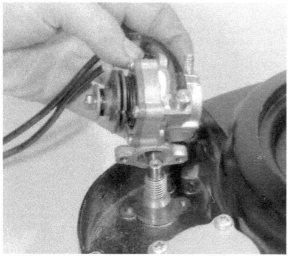

14.4 Pump is retained by two screws

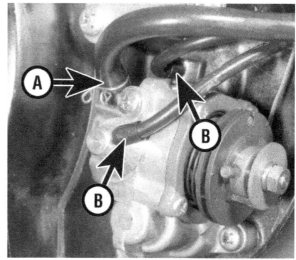

14.6 A: oil feed pipe from tank, B: oil delivery pipes

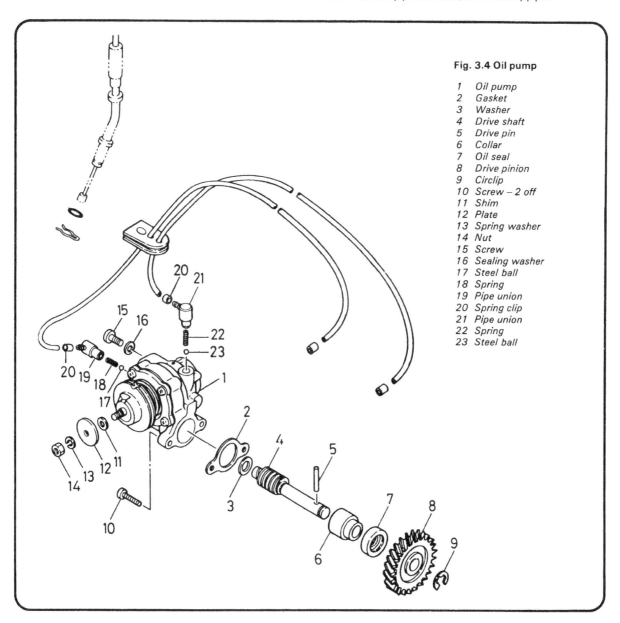

Fig. 3.4 Oil pump

1 Oil pump
2 Gasket
3 Washer
4 Drive shaft
5 Drive pin
6 Collar
7 Oil seal
8 Drive pinion
9 Circlip
10 Screw – 2 off
11 Shim
12 Plate
13 Spring washer
14 Nut
15 Screw
16 Sealing washer
17 Steel ball
18 Spring
19 Pipe union
20 Spring clip
21 Pipe union
22 Spring
23 Steel ball

15 Bleeding the oil pump

1 It is necessary to bleed the oil pump every time the main feed pipe from the oil tank is removed and replaced. This is because air will be trapped in the oil line, no matter what care is taken when the pipe is removed.

2 Check that the oil pipe is connected correctly, with the retaining clip in position. Then remove the cross-head screw in the outer face of the pump body with the fibre washer beneath the head. This is the oil bleed screw.

3 Check that the oil tank is topped up to the correct level, then place a container below the oil bleed hole to collect the oil that is expelled as the pump is bled. Allow the oil to trickle out of the bleed hole, checking for air bubbles. The bubbles should eventually disappear as the air is displaced by fresh oil. When clear of air, refit the bleed screw. DO NOT replace the front portion of the crankcase cover until the pump setting has been checked, as described in the next Section.

4 Note also that it will be necessary to ensure that the oil delivery pipes are primed if these have been disturbed. Unless this is checked the engine will be starved of oil until the pipes fill. The procedure required to avoid this is to start the engine and allow it to idle for a few minutes whilst holding the pump pulley in its fully open position by pulling the pump cable. The excess oil will make the exhausts smoke heavily for a while, indicating that the pump is delivering oil to the engine.

15.1 Screw is removed to facilitate pump bleeding

16 Checking the oil pump and throttle cable settings

1 As has been mentioned previously, the oil pump is controlled by the throttle twistgrip, and it is important that the pump lever is kept synchronized with the carburettor throttle valve. It should be noted that the oil pump setting must always be checked whenever the throttle cable is altered, as each adjustment will affect the relationship with the other.

2 Check that the carburettors are correctly adjusted and synchronized as described earlier in this Chapter, then set the throttle cable adjuster to give free play. Hold this setting, and secure the adjuster locknut.

3 The pump pulley carries an alignment marks or marks which should align with the pump plunger pin at full throttle. The mark used carries an alignment mark or marks which should align with the pump plunger pin at full throttle. The mark used varies according to the model and whether it has been modified to reduce oil consumption, as described below.

4 RD250/350 LC, 1980 models, prior to engine numbers 4L0 100101 (350) or 4L1 100101 (250). Use the raised rectangular mark on the edge of the pump pulley, about 8mm from the drilling for the cable nipple. Where modifications have been made to reduce oil consumption, a second mark should have been made about half way between the nipple drilling and the original mark. Where the second mark has been added, this should be used, noting that the minimum pump stroke should also be reduced from 0.20 – 0.25 mm to 0.15 mm.

5 RD250/350 LC, 1981 models, Engine numbers 4L0 100101 (350) or 4L1 100101 (250) onwards. The pump pulley carries two alignment marks, a rectangular mark about 8 mm from the cable nipple drilling and a round mark about halfway between the two. Check the pulley identification mark. If it is 1M1, use the rectangular mark for all models. Where a 4L1 mark is found, use the rectangular mark for 250 cc machines and the round mark on 350 cc models.

6 Having worked out the appropriate alignment mark, open the throttle twistgrip and check that the mark coincides with the pump plunger pin. Where adjustment is required it can be made using the in-line adjuster above the right-hand outer cover.

7 The pump's minimum stroke adjustment should be checked next. Start the engine and allow it to idle. Observe the front end of the pump unit, where it will be noticed that the pump adjustment plate moves in and out. When the plate is out to its

16.3 Check pump synchronisation at full throttle

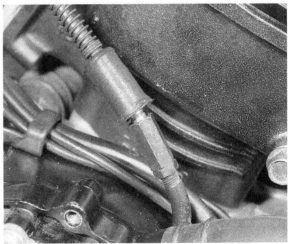

16.6 Pump cable adjuster

fullest extent, stop the engine and measure the gap between the plate and the raised boss of the pump pulley using feeler gauges. Do not force the feeler gauge into the gap – it should be a light sliding fit. Make a note of the reading, then repeat the procedure several times. The largest gap is indicative that the pump is at its minimum stroke position. If the pump is set up correctly, the gap found should be 0.20 – 0.25 mm (0.008 – 0.010 in) on unmodified 1980 models. In the case of the modified 1980 machines and all 1981 versions, engine numbers 4L1 100101 (250 cc) or 4L0 100101 (350 cc) onwards, the setting is reduced to 0.15 mm (0.006 in).

8   If the stroke setting is incorrect, remove the nut which secures the adjustment plate and add or subtract shims as required. If necessary, these can be purchased from Yamaha dealers in thicknesses of 0.3 and 0.5 mm (0.0118 and 0.0197 in). In practice, the pump is unlikely to require frequent adjustment.

17 Reed valve induction system: mode of operation

1   Of the various systems of controlling the induction cycle of a two-stroke engine, Yamaha has chosen to adopt the reed valve, a device which permits precise control of the incoming mixture, allowing more favourable port timing to give improved torque and power outputs. The reed valve assembly comprises a wedge-shaped die-cast aluminium alloy valve case mounted in the inlet tract. The valve case has rectangular ports which are closed off by flexible stainless steel reeds. The reads seal against a heat and oil resistant synthetic rubber gasket which is bonded to the valve case. A special shaped valve stopper, made from cold rolled stainless steel plate, controls the extent of movement of the valve reeds.

2   As the piston ascends in the cylinder, a partial vacuum is formed beneath the cylinder in the crankcase. This allows atmospheric pressure to force the valve open, and a fresh charge of petrol/air mixture flows past the valve and into the crankcase. As the pressure differential becomes equalised, the valves close, and the incoming charge is then trapped. The charge of mixture in the cylinder is by this time fully compressed, and ignition takes place driving the piston downwards. The descending piston eventually uncovers the exhaust port, and the hot exhaust gases, still under a certain amount of pressure, are discharged into the exhaust system. At this stage, the reed valve, in conjunction with the 7th, or auxiliary scavenging port, performs a secndary function; as the hot exhaust gases rush out of the exhaust port, a momentary depression is created in the cylinder, this allows the valve to open once more, but this time the incoming mixture enters directly into the cylinder via the 7th port and completes the expulsion of the now inert burnt gases. This ensures that the cylinder is filled with the maximum possible combustion mixture. The charge of combustion mixture which has been compressed in the crankcase is released into the cylinder via the transfer ports, and the piston again ascends to close the various ports and begin compression. The reed valves open once more as another partial vacuum is created in the crankcase, and the cycle of induction thus repeats. It will be noted that no direct mechanical operation of the valve takes place, the pressure differential being the sole controlling factor.

18 Reed valves: removal, examination and renovation

1   The reed valve assembly is a precision component, and as such should not be dismantled unnecessarily. The valves are located in the inlet tract, covered by the carburettor flanges.

2   Remove the carburettors as described in Section 6 of this Chapter thus exposing the four cross-headed screws retaining the adaptors and the reed valve assemblies to the cylinders. After removing these screws, the assemblies can be carefully lifted away.

16.7 Measure pump stroke using feeler gauges

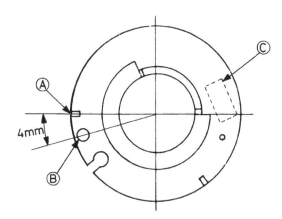

Fig. 3.5 Oil pump pulley alignment marks

A   Standard alignment mark for 1980 models; RD250 LC up to Engine No 4L1 100101 and RD350LC up to Engine No 4L0 100101

B   Modified alignment mark for the above models (see text)

C   Pulley identification mark for 1981 models; RD250 LC from 4L1 100101 and RD350LC from 4L0 100101 onwards. If pulley is marked 1M1 use alignment mark A. If pulley is marked 4L1 use mark A for 250cc models, and mark B for 350cc models

3   The valves can now be washed in clean petrol to facilitate further examination. They should be handled with great care, and on no account dropped. The stainless steel reeds should be inspected for signs of cracking or fatigue, and if suspect, should be renewed. Remember that any part of the assembly which breaks off in service will almost certainly be drawn into the engine, causing extensive damage.

4   Note the position of the reed in relation to the Neoprene bonded gasket against which it seats. It should be flush to form an effective seal. If further dismantling is deemed necessary proceed as follows.

5   Remove the two cross-head screws securing the valve stopper and reed to the case. Handle the reed carefully, avoiding bending, and note from which side it was removed. All components should be replaced in their original positions. A small cutout on the lower left-hand corner of the reed can be used to assist in relocation. Lay the reed carefully to one side if it is to be re-used. Examine the neoprene seating face which, is defective, will necessitate the replacement of the complete alloy case, to which it is heat-bonded.

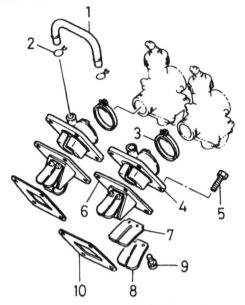

Fig. 3.6 Reed valve assembly

1 Interconnecting pipe
2 Pipe clip – 2 off
3 Hose clamp
4 Reed valve mounting
5 Bolt – 4 off
6 Reed valve
7 Reed valve petal
8 Reed valve stopper
9 Screw and washer – 2 off
10 Gasket

6   Reassembly is a direct reversal of the dismantling process. Clean all parts thoroughly, but gently, before refitting. A thread locking compound, such as Loctite, must be applied to the two cross-headed screws, which should be tightened progressively to avoid warping the reed or stopper. Do not omit the locking compound, as the screws retain a component which vibrates many times each second and consequently are prone to loosening if assembled incorrectly.

7   The assembly should now be checked before refitting. The dimensions between the inner edge of the valve stopper and the top edge of the valve case is important as it controls the movement of the reed. If smaller than specified, performance will be impaired. More seriously, if larger than specified, the reed may fracture. The nominal setting is 9.0 mm (0.36 in). If the measurement is more than 0.4 mm (0.016 in) either side of the nominal figure, the stopper plate should be renewed. Check the valve case mating surfaces for warping, renewing or resurfacing as necessary.

8   Assembly is a direct reversal of the removal sequence. Fit new gaskets between the cylinder and valve assembly. Tighten the Allen screws evenly to avoid warping. Fit the carburettors as described in Section 6.

18.2a Remove inlet adaptors to expose reed valves ...

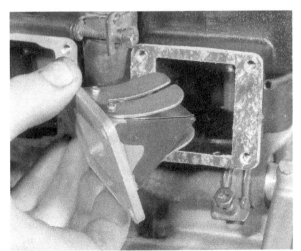

18.2b ... which can be removed for examination

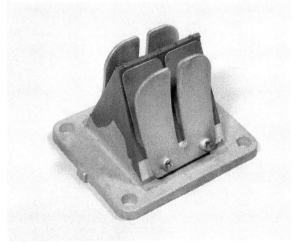

18.3 The reed valve assembly

Chapter 4 Ignition system

Contents

General description 1
Electronic ignition system: principles of operation 2
Electronic ignition system: testing and maintenance 3
Wiring, connectors and switches: checking 4
Pulser coil: testing 5
Exciter coils (source coils): testing 6
Ignition coil: testing 7
CDI unit: testing 8
Checking the ignition timing 9
Spark plugs: checking and resetting the gaps 10

Specifications

Ignition system

Type	Capacitor discharge (CDI)
Make	Nippon Denso
Model	VCC27
Voltage	12 volt
Pulser coil resistance:	
White/red to Black	87 ohms $\pm$ 10% @ 20°C (68°F)
Source coil resistance:	
Brown to Black	271 ohms $\pm$ 10% @ 20°C (68°F)
Brown to Red	5.1 ohms $\pm$ 10% @ 20°C (68°F)

Ignition coil

Make	Nippon Denso
Model	129700-027/
Minimum spark gap	6.0 mm (0.24 in)
Primary winding resistance	0.33 ohms $\pm$ 10% @ 20°C (68°F)
Secondary winding resistance	3.5 k ohms $\pm$ 20% @ 20°C (68°F)

CDI unit

Make	Nippon Denso
Model	0700000-072/

Spark plug

Make	NGK
Type	B8ES
Gap	0.7 - 0.8 mm (0.028 - 0.032 in)

1 General description

1   The Yamaha RD250 and RD350 LC models are equipped with fully electronic capacitor discharge ignition (CDI) systems. This arrangement provides a more powerful and accurate ignition and can be considered almost maintenance-free.

2   The most important advantage of electronic ignition, is that it removes all mechanical components from the system, the spark being triggered electronically by a pulser coil and magnet rather than by a contact breaker assembly. Because there are no contact breakers to wear, the owner is freed from the task of periodically adjusting or renewing them. Once the electronic system has been set up, it need not be attended to unless it has been disturbed in the course of dismantling or failure occurs in the electronic components in the system.

3   The ignition exciter, or source, coil assembly is housed within the flywheel generator, and provides the power supply for the external CDI unit, which is mounted beneath the right-hand side panel. The system is triggered by a magnet, in-

corporated in the outer face of the generator rotor, acting upon a pickup coil, known as a pulser, which is outrigged from the stator. The spark plugs are supplied from a single ignition coil, firing in both cylinders simultaneously. This system is known as the 'spare spark' system, as only one cylinder can fire, leaving one spark wasted, or 'spare'.

2 Electronic ignition system: principles of operation

1 Energy for the ignition system is drawn from the exciter coil. This is mounted on the generator stator, and is integral with the normal alternator windings. It is a two-stage arrangement, having low speed and high speed windings. The low speed windings produce a high output voltage at low engine speeds, this voltage dropping off as the engine builds up speed. The high speed windings, on the other hand, produce little energy at

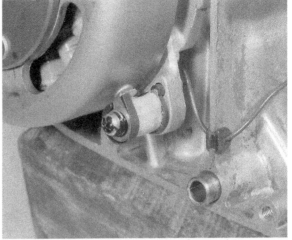

1.3 Pulser coil is mounted on extension of stator plate

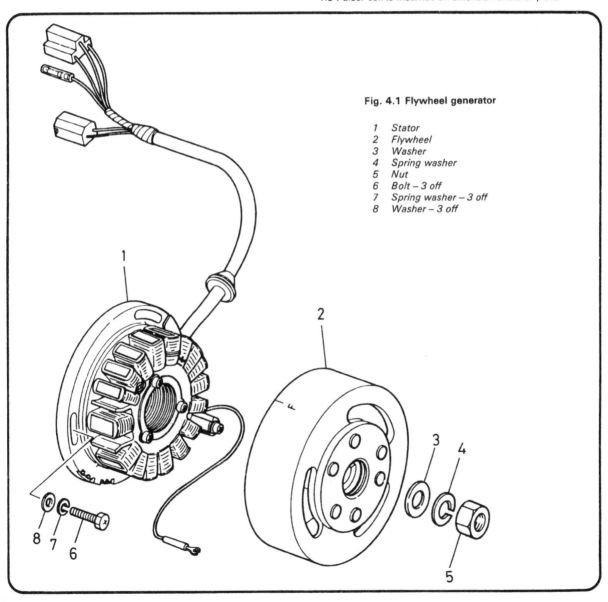

Fig. 4.1 Flywheel generator

1 Stator
2 Flywheel
3 Washer
4 Spring washer
5 Nut
6 Bolt – 3 off
7 Spring washer – 3 off
8 Washer – 3 off

low engine speeds, but the output voltage rises along with engine speed. The two outputs are combined, offsetting each other to give a fairly constant output voltage, this being the sum of the output of each set of windings.

2   The exciter coil assembly feeds the CDI unit, a sealed electronic assembly which forms the heart of the system. This unit contains, amongst other things, a capacitor and a thyristor, or silicon controlled rectifier (SCR). The capacitor is charged with the high voltage output from the exciter coil assembly. The thyrister, or SCR, is in effect an electronic switch. When signalled electrically by the pulser, it allows the capacitor to discharge through the primary windings of the ignition coil. This in turn induces a high tension pulse in the secondary windings, which is fed to the spark plugs.

3   The pulser, or pickup, comprises a small coil mounted outside the alternator rotor on a projection from the stator. A permanent magnet embedded in the flywheel rotor is arranged to pass beneath the pulser coil. As the magnet passes the pulser coil, a weak current is generated and it is this that is used to trigger the thyristor in the CDI unit.

3   Electronic ignition system: testing and maintenance

1   As stated earlier in this Chapter, the electronic ignition needs no regular maintenance once it has been set up and timed accurately. Occasional attention should, however, be directed at the various connections in the system, and these must be kept clean and secure. A failure in the ignition system is comparatively rare, and usually results in a complete loss of ignition. Usually, this will be traced to the CDI unit, and little can be done at the roadside to effect a repair. In the event of the CDI unit failing, it must be renewed as repair is not practicable.

2   If the CDI unit is thought to be at fault, it is recommended that it be removed and taken to a Yamaha dealer for testing. The dealer will have the use of diagnostic equipment, and will be able to test the unit accurately and quickly. Testing at home is less practical and cannot be guaranteed to be accurate. At best, it will enable the owner to establish which part of the system is at fault, although replacement of the defective part remains the only effective cure.

3   Care must be exercised when dealing with the CDI unit. Wrong connections could cause instant and irreparable damage to the unit, so this must be an important consideration when removing and replacing the unit. If the unit is to be disconnected and removed for testing, use a short length of insulated wire to short-circuit each pair of terminals to avoid any electric shock from residual energy in the capacitor.

4   For those owners possessing a multimeter and who are fully conversant with its use, a test sequence is given below. It is not recommended that the inexperienced attempt to test the system at home, as more damage could be sustained by the system, and an unpleasant shock experienced by the unwary operator. It should be appreciated that the CDI system is capable, in certain circumstances, of producing sufficient current output to be dangerous to the operator. It follows that an attitude to safety should be adopted similar to that applying when testing household mains electricity. Before performing any tests, the following point must be noted. Disconnect all the connectors in the ignition system, thus isolating the various components.

5   It will be noted that some of the resistance tests require a meter capable of reading in ohms rather than kilo ohms. Many of the cheaper multimeters are only capable of reading the latter, and thus are of limited use for accurate testing of low-resistance components. When testing the ignition system to isolate faults it is important to follow a logical sequence to avoid wasted effort and money. Use the accompanying flow charts as a guide, referring to the subsequent sections for more detailed information.

a) No spark is produced, or weak spark

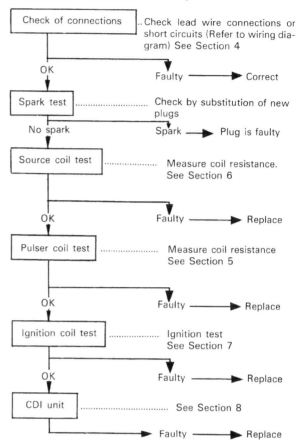

b) The engine starts but will not pick up speed

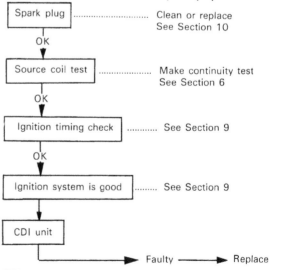

4   Wiring, connectors and switches: checking

1   The single most likely cause of ignition failure or partial failure is a broken, shorted or corroded connector, switch contact or wire. Tracing faults of this nature can often prove time consuming but does not require the use of sophisticated test equipment. Although in rare instances a particular fault will not be evident from a physical examination, it is worthwhile

Electrode gap check - use a wire type gauge for best results

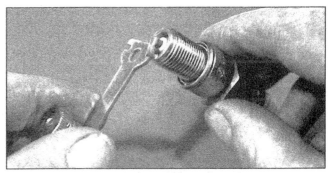

Electrode gap adjustment - bend the side electrode using the correct tool

Normal condition - A brown, tan or grey firing end indicates that the engine is in good condition and that the plug type is correct

Ash deposits - Light brown deposits encrusted on the electrodes and insulator, leading to misfire and hesitation. Caused by excessive amounts of oil in the combustion chamber or poor quality fuel/oil

Carbon fouling - Dry, black sooty deposits leading to misfire and weak spark. Caused by an over-rich fuel/air mixture, faulty choke operation or blocked air filter

Oil fouling - Wet oily deposits leading to misfire and weak spark. Caused by oil leakage past piston rings or valve guides (4-stroke engine), or excess lubricant (2-stroke engine)

Overheating - A blistered white insulator and glazed electrodes. Caused by ignition system fault, incorrect fuel, or cooling system fault

Worn plug - Worn electrodes will cause poor starting in damp or cold weather and will also waste fuel

checking the obvious points before resorting to expensive professional assistance.

2   Refer to the wiring diagram at the end of the book, trace and check each of the alternator output leads, connector blocks and all connections to the CDI unit, ignition switch and kill switch. Do not ignore the obvious; it can be very frustrating to spend a long time checking the ignition system, only to discover that the kill switch was set to the 'Off' position. The kill switch and some of the wiring connectors are relatively exposed and may have become contaminated by water. To eliminate this possibility, spray each one with a water dispersing aerosol such as WD40.

3   Check the wiring for breakages or chafing paying particular attention to the wiring in areas like the steering head where steering movement causes flexing. The plastic insulation may appear intact even if the internal conductor has broken. If a wire is suspected of having broken internally or shorted against the frame it should be checked using a multimeter as a continuity tester.

5   Pulser coil: testing

1   Trace the alternator output wiring back to the connector blocks beneath the seat. Separate the pulser lead connector (White/red lead) and the three-pin connector carrying the leads to the CDI unit (Red, Brown and Black leads).

2   Using an ohmmeter or multimeter set at the ohms x 10 scale, measure the resistance between the White/red pulser lead and the Black earth lead. The pulser can be considered serviceable if the following resistance figures are obtained.

Pulser coil resistance
    White/red to black   87 ohms ± 10% @ 20°C (68°F)

6   Exciter coils (source coils): testing

1   The exciter coils, or source coils are connected to the CDI unit via three wires (Red, Brown and Black) and pass through a three-pin connector located beneath the dual seat. Separate the connector and make the following test on the alternator side of the wiring.

2   Set the meter on the ohms x 1 scale and measure the resistance between the Red and Brown leads (high speed windings). Next, with the meter set on ohms x 10, measure the resistance between the Brown lead and the Black (earth) lead. The readings should compare with those shown below.

Exciter (source) coils – resistance
    Red to Brown leads   5.1 ohms ± 10% @ 20°C (68°F)
    Brown to Black leads   271 ohms ± 10% @ 20°C (68°F)

7   Ignition coil: testing

1   If the ignition coil is suspected of having failed it can be tested by measuring the resistance of its primary and secondary windings. The test can be performed with the coil in place on the frame, having first disconnected the high tension leads at the spark plugs and the low tension lead at the connector block.

2   Connect one of the meter probes to earth and the other to the low tension lead at the connector block. This will give a resistance reading for the primary windings and should be within the limits shown below.

Primary winding resistance
    0.33 ohms ± 20% @ 20°C (68°F)

3   Next, connect a meter probe lead to each of the high tension leads to measure the secondary winding resistance and compare the reading obtained with that shown below.

Secondary winding resistance
    3.5 kilo ohms ± 30% @ 20°C (68°F)

4   If either of the values obtained differs markedly from the specified resistances it is likely that the coil is defective. It is recommended that the suspect coil is taken to a Yamaha dealer who can then verify the coil's condition and supply a replacement unit where necessary. The coil is a sealed unit and therefore cannot be repaired.

8   CDI unit: testing

1   If the tests shown in the preceding Sections have failed to isolate the cause of an ignition fault it is likely that the CDI unit is itself faulty. Whilst it is normally possible to check this by making resistance measurements across the various terminals, Yamaha do not supply the necessary data for the LC models. It follows that it will be necessary to enlist the help of a Yamaha dealer who will be able to check the operation of the unit by substituting a sound item.

7.1a Dismantle plastic shroud to expose coil

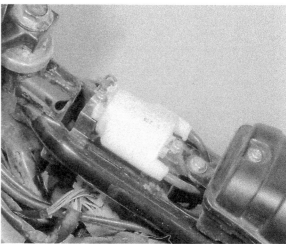

7.1b Coil is mounted on frame forward of the air cleaner

9   Checking the ignition timing

1   To check the timing with the required degree of accuracy it will be necessary to use a dial gauge and a stroboscopic timing lamp. A dial gauge can be purchased through Yamaha dealers as a set comprising the gauge, an adaptor and an extension needle as part number 90890 - 01252. A stroboscopic timing lamp or 'strobe' is also available as part number 90890 - 03109. Both items can also be purchased from independent suppliers, many of whom advertise regularly in the motorcycle press. These tools are essential, and if they are not available the work must be entrusted to a Yamaha dealer.

2   Start by removing the left-hand engine casing to expose the alternator rotor. This will necessitate removal of the gearchange linkage. Remove the left-hand spark plug and screw the dial gauge adaptor into the spark plug hole. Fit the dial gauge extension needle to the gauge, then fit the assembly into the adaptor. Rotate the crankshaft by turning the alternator rotor. As the piston approaches top dead centre (TDC) the gauge reading will increase, stopping momentarily as TDC is passed and then decreasing as the piston begins to descend. Rock the crankshaft to and fro until the exact position of TDC is found, then set the gauge to read zero at this point. Move the rotor back and forth a few times to make sure that the needle does not move past zero.

3   Observing the gauge, turn the rotor clockwise until a reading of 3 or 4 mm is shown, then slowly move it anticlockwise until the piston is 2.0 mm (0.08 in) BTDC. Check the 'F' mark on the rotor in relation to the fixed index mark on the baseplate extension. If necessary, slacken the three baseplate bolts and move the baseplate until the marks coincide. The bolts can be slackened using an open ended spanner passed between the baseplate and the rotor edge. When the timing is correct, tighten the bolts and re-check the setting as described above.

4   Remove the dial gauge assembly and refit the spark plug. Connect the timing light as directed by the manufacturer's instructions, then start the engine. Direct the light at the 'F' mark on the rotor. The timing lamp will flash each time the plug sparks and will make the timing mark appear to freeze at that point. If all is well, the 'F' mark should coincide with the fixed index mark at 2000 rpm. In the unlikely event that there is a discrepancy, note where the 'F' mark does appear, stop the engine, and inscribe a new fixed index line in the appropriate place. The new line should be used for subsequent timing checks.

10 Spark plugs: checking and resetting the gaps

1   The Yamaha LC models are equipped with two NGK B8ES spark plugs as standard fitments. Certain operating conditions may indicate the need for a change in spark plug grade, but generally the type recommended by thhe manufacturer gives

9.3 'F' mark should align with fixed index mark

the best all round service.

2   Check the gap of the plug points every monthly or 1000 mile service. To reset the gap, bend the outer electrode to bring it closer to, or further away from the central electrode until a 0.7 mm (0.028 in) feeler gauge can be inserted. Never bend the centre electrode or the insulator will crack, causing engine damage if the particles fall into the cylinder whilst the engine is running.

3   With some experience, the condition of the spark plug electrodes and insulator can be used as a reliable guide to engine operating conditions. See the accompanying series of colour photographs.

4   Always carry spare spark plugs of the recommended grade. In the rare event of plug failure, this will enable the engine to be restarted.

5   Beware of over-tightening the spark plugs, otherwise there is risk of stripping the threads from the aluminium alloy cylinder heads. The plugs should be sufficiently tight to seat firmly on their copper sealing washers, and no more. Use a spanner which is a good fit to prevent the spanner from slipping and breaking the insulator.

6   If the threads in the cylinder head strip as a result of over-tightening the spark plugs, it is possible to reclaim the head by the use of a Helicoil thread insert. This is a cheap and convenient method of replacing the threads; most motorcycle dealers operate a service of this nature at an economic price.

7   Make sure the plug insulating caps are a good fit and have their rubber seals. They should also be kept clean to prevent tracking. These caps contain the suppressors that eliminate both radio and TV interference.

Chapter 5 Frame and forks

Contents

General description 1
Front forks: removal and replacement 2
Steering head assembly: dismantling and reassembly 3
Front forks: dismantling and reassembly 4
Front forks: examination and renovation 5
Steering head bearings: examination and renovation 6
Steering lock 7
Frame: examination and renovation 8
Rear suspension assembly: general information 9
Rear sub-frame: examination and renovation 10
Rear suspension unit: examination 11

Centre stand: examination and maintenance 12
Prop stand: examination 13
Footrests: examination and renovation 14
Rear brake pedal: examination and renovation 15
Speedometer and tachometer heads: removal and
replacement 16
Speedometer and tachometer drive cables: examination
and maintenance 17
Speedometer and tachometer drives: location and
examination 18

Specifications

Frame

Type .... Welded tubular steel

Front forks

Type .... Oil damped telescopic
Spring free length .... 497.8 mm (19.6 in)
Fork oil grade .... SAE 10W/30 engine oil
Fork oil capacity (per leg)* .... 140 ± 2.5 cc (4.93 ± 0.098 Imp fl oz)
Fork oil level (below top of stanchion) .... 195 ± 5.5 mm (7.68 ± 0.22 in)

No fork leg oil capacitites are available other than those given. It is suggested that the quantity per leg be reduced by about 10 cc when refilling forks which have been drained only and not dismantled.

Rear suspension

Type .... Cantilever

Rear suspension unit

Type .... Single gas/oil and coil spring
Travel .... 55 mm (2.16 in)
Wheel travel .... 110 mm (4.33 in)
Spring free length .... 216 mm (8.50 in)
Gas pressure .... 15 kg/cm² (213 psi)

---

1 General description

1 The Yamaha RD250/350 LC models use conventional welded tubular steel frames. The front wheel is supported by oil-damped coil spring telescopic suspension units. Rear suspension is by a cantilever assembly, pivoting on plain bushes, supported by a single suspension unit. The unit is coil sprung with nitrogen under high pressure as a supplementary springing medium. Damping is hydraulic.

2 Front forks: removal and replacement

1 It is unlikely that the front forks will need to be removed from the frame together with the fork yoke and steering stem. Although feasible, this method is both unwieldy and time consuming. It is far easier to remove the individual fork legs, then the steering head assembly, where necessary, as described in Section 3 of this Chapter.

2 Place the machine on its centre stand. Make sure that it is standing securely on a flat surface, then place blocks beneath the front of the engine to raise the front wheel clear of the ground. Unscrew the knurled retaining ring which secures the speedometer drive cable to its gearbox on the front wheel. Pull the cable clear and lodge it clear of the wheel and forks.

3 On the 350 cc model, remove the four mudguard retaining bolts and lift the mudguard clear of the forks. In the case of the 250cc model the mudguard can be removed now or left until the wheel has been removed, if desired.

4 Straighten and remove the split pin which secures the wheel spindle nut. Slacken and remove the nut, then support the wheel while the spindle is displaced and removed. If necessary, tap the spindle through but take care not to damage its threads. Once the spindle has been removed the wheel can be lowered to the ground. In the case of 250 cc machines the wheel can now be lifted clear. On 350 cc machines, turn both brake calipers outwards so that they do not foul the wheel rim and tyre, then remove the wheel.

5 It is good practice to fit a wooden wedge between the brake pads to prevent the caliper pistons from being displaced should the brake lever be squeezed while the wheel is removed. The brake caliper (or calipers, 350 cc model) should be released from the fork legs by removing the two mounting bolts. Lift the caliper(s) away, tying them clear of the forks. Do not allow the caliper(s) to hang from the hydraulic hoses.

6 If the forks are to be dismantled it is better to release the fork top plugs at this stage. It can be done with the forks removed, as shown in the photograph accompanying this Section, but for convenience they should be removed while the fork stanchions are secured in the yokes. Start by prising off the black plastic caps which cover the stanchion tops. This will reveal the fork plugs which will be found to be retained by a wire internal circlip.

7 Try to obtain some assistance at this stage. The plugs are removed by pushing them down against fork spring pressure and the easiest method is to use a cross-point screwdriver in the central depression in the cap. While the cap is held down it will be necessary for the second person to dislodge the retaining clip with a small electrical screwdriver. Once this has been removed the plug can be slowly released and will be pushed out of the stanchion. This operation can be accomplished unaided, but it is best to have two pairs of hands if possible.

8 Slacken the pinch bolts in the upper and lower fork yokes and slacken the screws which clamp the headlamp brackets to the fork stanchions. The latter brackets should be supported while the stanchions are twisted and pulled downwards to free them.

9 The fork legs are refitted by reversing the removal sequence described above. Slide each fork leg into position, ensuring that the tops of each stanchion are level. Tighten the upper and lower pinch bolts. On all models, if the steering head assembly has been dismantled, it is important to ensure that the forks are correctly aligned when refitted. To this end, assemble the forks loosely, refit the front wheel, then bounce the forks a few times so that the various components assume their correct relative positions. Tighten the clamp bolts, working from the wheel spindle upwards.

10 Note the following torque settings when assembling the forks and calipers.

Fork yoke pinch bolts	2.4 kgf m (17.2 lbf ft)
Wheel spindle nut	7.4 kgf m (53.0 lbf ft)
Caliper mounting bolts	3.5 kgf m (25.0 lbf ft)

2.2 Release knurled ring and detach speedometer cable

2.3 Mudguard is secured by four bolts

2.4a Remove split pin and wheel spindle nut ...

2.4b ... then withdraw spindle and remove wheel

2.5a Remove caliper mounting bolts ...

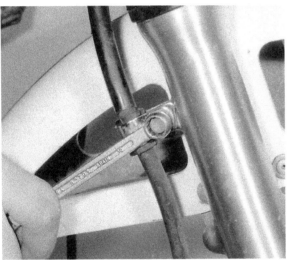

2.5b ... and hose retainer to free hydraulics

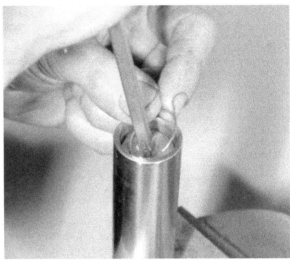

2.7 Depress fork top plug and release retaining clip

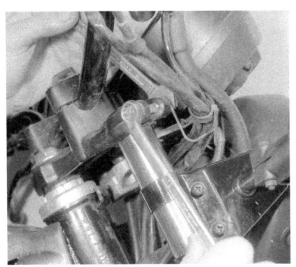

2.8a Slacken upper pinch bolt and nut ...

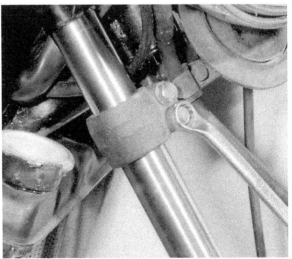

2.8b ... followed by lower pinch bolts and ...

2.8c ... then free the headlamp bracket assembly

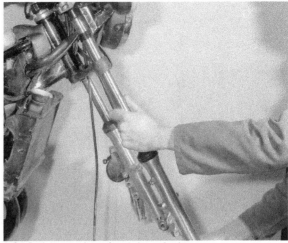

2.8d Fork leg can be twisted clear of yokes

3   Steering head assembly: dismantling and reassembly

1   Commence dismantling by removing the front fork legs as described in Section 4. The petrol tank should be removed to prevent accidental damage to its paint finish, or alternatively it can be covered with a thick blanket or similar padding.
2   Remove the headlamp unit from its shell after releasing the two screws which retain it. Disconnect the headlamp leads and place the unit to one side. Trace and disconnect the wiring from the handlebar switches at their connectors inside the headlamp shell. Unscrew the knurled rings which secure the speedometer and tachometer drive cables to the underside of the instrument heads and pull the cables clear of the steering head. Slacken off the clutch cable adjuster and free it from the operating lever.
3   Remove the two bolts which retain the instrument panel assembly to the top yoke. The instrument panel can now be pulled away from the fork yokes together with the headlamp shell and brackets and the front indicator lamps, freeing the wiring connectors as the assembly is pulled clear. Remove the front brake hose union from the bottom yoke and also the horn and its mounting bracket.
4   Prise off the black plastic caps which cover the handlebar clamp screws. The screws should be released to allow the clamp halves to be removed. The handlebar assembly can now

be lifted clear of the top yoke and lodged across the top of the frame, or rested on the fuel tank.
5   Before proceeding further, it should be noted that the steering head bearings are of the uncaged ball type, and these will drop free as the steering stem and lower yoke are released. The steel balls will drop free if left unattended, so some provision must be made to catch them. It is a good idea to place some clean cloth under the steering head area to contain any errant balls.
6   Remove the crown bolt from the centre of the top yoke an lift the yoke from position. Using a C-spanner, slacken and remove the steering stem nut, whilst supporting the lower yoke. With the nut removed, carefully lower the yoke and steering stem catching the balls from the lower race as they drop free. It will be noted that there are nineteen $\frac{1}{4}$ in balls in each race.
7   The steering head assembly should be reassembled in the reverse order of that given for dismantling. When fitting the steel balls to their races, they can be held in place with grease. Check that the correct number is fitted to each bearing race. When fitting the steering stem nut, it must be adjusted so that all perceptible free play is taken up, but no more. It is easy to damage the head races by overtightening. When correctly adjusted, it should be possible to move the steering from lock to lock with the lightest pressure on the handlebar end. Final adjustment can be made after reassembly, by slackening the top bolt and adjusting the steering stem nut as required.

3.2 Disconnect drive cables at instrument heads

3.3 Instrument panel is retained by two bolts

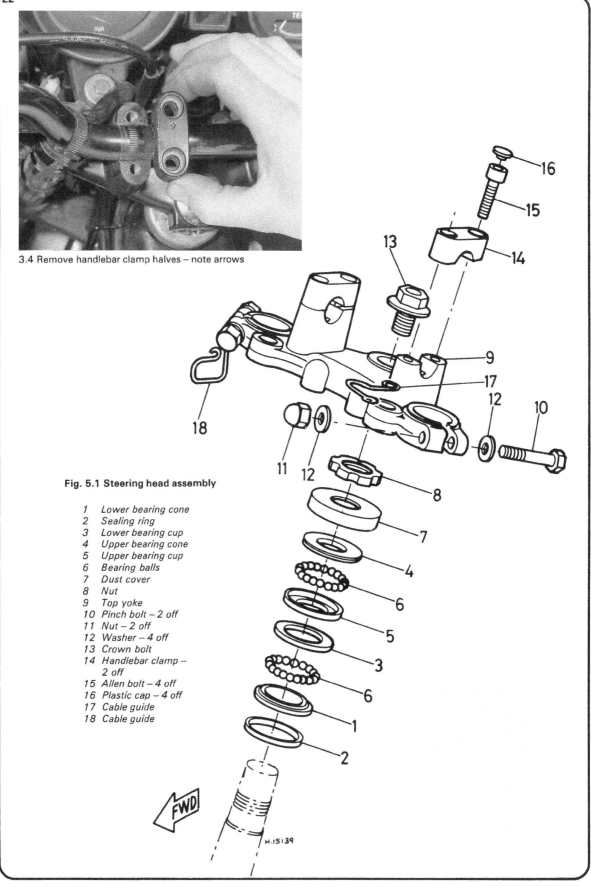

3.4 Remove handlebar clamp halves – note arrows

Fig. 5.1 Steering head assembly

1   Lower bearing cone
2   Sealing ring
3   Lower bearing cup
4   Upper bearing cone
5   Upper bearing cup
6   Bearing balls
7   Dust cover
8   Nut
9   Top yoke
10 Pinch bolt – 2 off
11 Nut – 2 off
12 Washer – 4 off
13 Crown bolt
14 Handlebar clamp –
    2 off
15 Allen bolt – 4 off
16 Plastic cap – 4 off
17 Cable guide
18 Cable guide

H.15139

4 Front forks: dismantling and reassembly

1   Having removed the forks from the yokes as described in Section 2 withdraw the fork springs and invert each leg over a drain tray until the damping oil has emptied. It is assumed that the fork top plugs were removed as described prior to their removal from the fork yokes. If the plugs are still in position they can be removed by clamping the stanchion in a vice fitted with soft jaws. Take care not to crush or score the stanchion. Depress the plug with a cross-point screwdriver and prise out the wire circlip which retains it. Release the pressure on the plug, allowing it to be displaced by the fork spring. The fork leg can now be removed and drained as described above.

2   Once the oil has been drained, slacken the bolt which passes up through the bottom of the lower leg and into the damper rod. It is quite likely that the damper rod will tend to rotate in the lower leg and thus impede the removal of the bolt. If this problem arises, clamp the assembly in a vice using soft jaws to hold the lower leg by the caliper mounting lugs. Obtain a length of wooden dowel about $\frac{1}{2}$ in in diameter and form a taper on one end.

3   Pass the dowel down the stanchion, having first withdrawn the fork spring. Push the dowel hard against the head of the damper rod to lock it in position whilst an assistant slackens the retaining bolt. If the dowel proves difficult to hold a self-grip wrench or similar can be used to obtain sufficient leverage and pressure. Once the retaining bolt has been removed slide off the dust seal and withdraw the stanchion assembly from the lower leg.

4   Invert the stanchion to release the damper rod and rebound spring, placing these components to one side to await examination. Unless new oil seals have been fitted recently they should be renewed as a matter of course to prevent failure at a later date. Lever out the wire retaining clip using a small screwdriver.

5   The oil seal can be levered out of its recess in the top of the lower leg and the spacer beneath it tipped out. Do not omit the latter when fitting new oil seals. The oil seal lip should be lubricated prior to installation, and then tapped into place using a large tubular drift to ensure that it enters the leg squarely. A large socket is ideal for this purpose. Remember to fit the retaining clip, ensuring that it locates correctly.

6   After the fork components have been examined for wear as described in Section 5, reassembly can commence. Make sure that all components are fitted to the leg from which they were removed. Slide the damper rod and rebound spring into the stanchion, then lubricate the stanchion surface with oil. Slide it into the lower leg making sure that the seal lip does not become distorted or damaged. Lock the damper rod, then fit the retaining bolt and its sealing washer, tightening it securely. Refit the dust seal over the stanchion and work it over the locating groove around the lower leg.

7   Slide the fork spring into position in the stanchion noting that the tightest coils must be uppermost. Top up each leg with SAE 10W/30 engine oil. Each leg should contain 140 ± 2.5 cc of oil (4.93 ± 0.098 Imp fl oz). The oil content is best measured by checking its level below the top of the stanchion using a dip stick. The correct level is 195 ± 5.5 mm (7.68 ± 0.22 in).

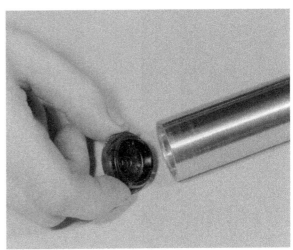

4.1a Fork top is covered by black plastic cap

4.1b Remove top plug, if still in position ...

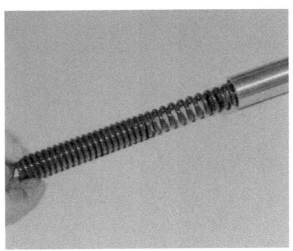

4.1c ... and withdraw the fork spring

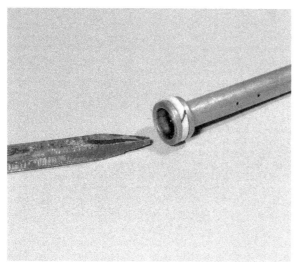

4.3a Tapered dowel or rod will prevent damper rotation ...

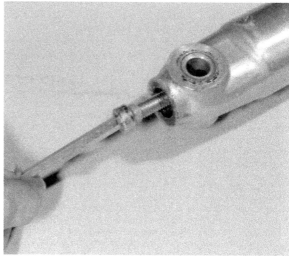

4.3b ... whilst bolt is unscrewed from fork leg

4.3c Slide off the rubber dust seal ...

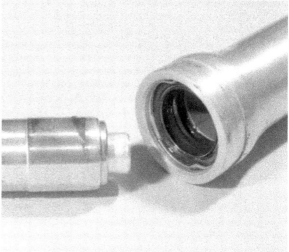

4.3d ... and withdraw the stanchion

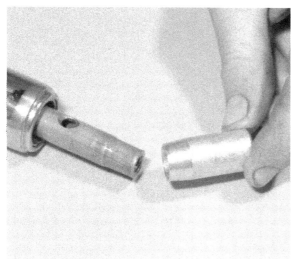

4.4a Remove damper seat ...

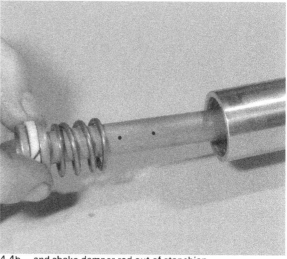

4.4b ... and shake damper rod out of stanchion

4.4c Fork oil seal is retained by wire clip

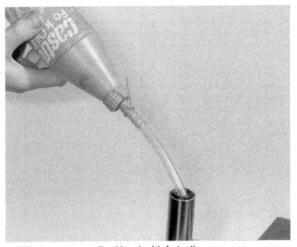

4.7 Top up to prescribed level with fork oil

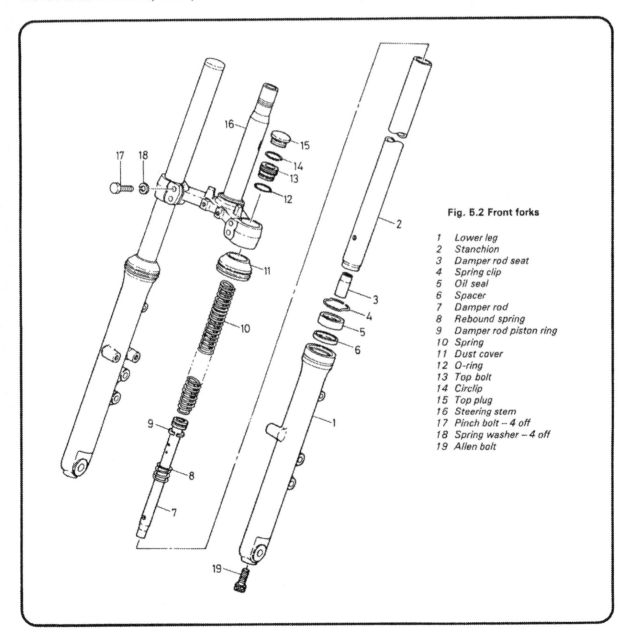

Fig. 5.2 Front forks

1   Lower leg
2   Stanchion
3   Damper rod seat
4   Spring clip
5   Oil seal
6   Spacer
7   Damper rod
8   Rebound spring
9   Damper rod piston ring
10 Spring
11 Dust cover
12 O-ring
13 Top bolt
14 Circlip
15 Top plug
16 Steering stem
17 Pinch bolt – 4 off
18 Spring washer – 4 off
19 Allen bolt

5   Front forks: examination and renovation

1   The parts most liable to wear over an extended period of
service are the internal surfaces of the lower leg and the outer
surfaces of the fork stanchion or tube. If there is excessive play
between these two parts, they must be renewed as a complete
unit. Check the fork tube for scoring over the length which
enters the oil seal. Bad scoring here will damage the oil seal and
lead to fluid leakage.
2   It is advisable to renew the oil seals when the forks are
dismantled, even if they appear to be in good condition. This will
save a strip down of the forks at a later date if oil leakage
occurs. The oil seal in the top of each lower fork leg is retained
by an internal C-ring which can be prised out of position with a
small screwdriver. Check that the dust excluder rubbers are not
split or worn where they bear on the fork tube. A worn excluder
will allow the ingress of dust and water which will damage the
oil seal and eventually cause wear of the fork tube.
3   It is not generally possible to straighten forks which have
been badly damaged in an accident, particularly when the
correct jigs are not available. It is always best to err on the side
of safety and fit new ones, especially since there is no easy
means to detect whether the forks have been over stressed or
metal fatigued. Fork stanchions (tubes) can be checked, after
removal from the lower legs by rolling them on a dead slat
surface. Any misalignment will be immediately obvious.
4   The fork springs will take a permanent set after considerable usage and will need renewal if the fork action becomes
spongy. The service limit for the total free length of each spring
is as follows.

Fork spring free length   497.8 mm (19.6 in)

6   Steering head bearings: examination and renovation

1   Before commencing reassembly of the forks, examine the
steering head races. The ball bearing tracks of the respective
cup and cone bearings should be polished and free from
indentations, cracks or pitting. If signs of wear are evident, the
cups and cones must be renewed. In order for the straight line
steering on any motorcycle to be consistently good, the steering
head bearings must be absolutely perfect. Even the smallest
amount of wear on the cups and cones may cause steering
wobble at high speeds and judder during heavy front wheel
braking. The cups and cones are an interference fit on their
respective seatings and can be tapped from position with a
suitable drift.
2   Ball bearings are relatively cheap. If the originals are
marked or discoloured they must be renewed. To hold the steel
balls in place during reassembly of the fork yokes, pack the
bearings with grease. Each race is fitted with nineteen steel
balls. Although each race has room for an extra steel ball it must
not be fitted. The gap allows the bearings to work correctly,
stopping them skidding and accelerating the rate of wear.

7   Steering lock

1   The steering lock is mounted forward of the steering head
lug where it is secured by a single bolt. When the steering is
turned fully to the right or left and the lock is operated the
mounting bolt is inaccessible and therefore unauthorised removal is prevented.
2   If the lock malfunctions repair is impracticable and a new
lock must be fitted. It follows that new keys must be acquired
at the same time.

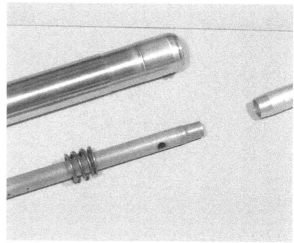

5.1a Check stanchion and damper parts for wear

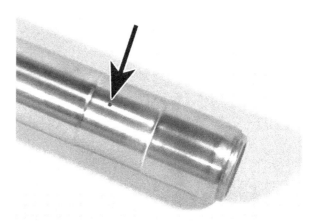

5.1b Ensure that small oil drilling is clear

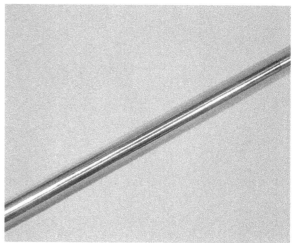

5.3 Stanchion is rolled on flat surface to check straightness

8 Frame: examination and renovation

1   The frame is unlikely to require attention unless accident damage has occurred. In some cases, replacement of the frame is the only satisfactory course of action if it is badly out of alignment. Only a few frame repair specialists have the jigs amd mandrels necessary for resetting the frame to the required standard of accuracy and even then there is no easy means of assessing to what extent the frame may have been over-stressed.

2   After the machine has covered a considerable mileage, it is advisable to examine the frame closely for signs of cracking or splitting at the welded joints. Rust can also cause weakness at these joints. Minor damage can be repaired by welding or brazing, depending on the extent and nature of the damage.

3   Remember that a frame which is out of alignment will cause handling problems and may even promote 'speed wob-bles'. If misalignment is suspected, as the result of an accident, it will be necessary to strip the machine completely so that the frame can be checked and, if necessary, renewed.

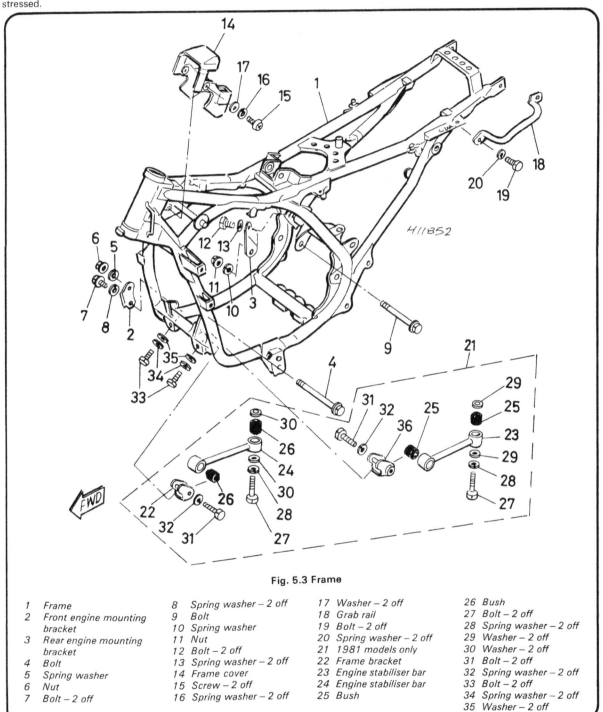

Fig. 5.3 Frame

1	Frame	8	Spring washer – 2 off	17	Washer – 2 off	26	Bush
2	Front engine mounting bracket	9	Bolt	18	Grab rail	27	Bolt – 2 off
3	Rear engine mounting bracket	10	Spring washer	19	Bolt – 2 off	28	Spring washer – 2 off
4	Bolt	11	Nut	20	Spring washer – 2 off	29	Washer – 2 off
5	Spring washer	12	Bolt – 2 off	21	1981 models only	30	Washer – 2 off
6	Nut	13	Spring washer – 2 off	22	Frame bracket	31	Bolt – 2 off
7	Bolt – 2 off	14	Frame cover	23	Engine stabiliser bar	32	Spring washer – 2 off
		15	Screw – 2 off	24	Engine stabiliser bar	33	Bolt – 2 off
		16	Spring washer – 2 off	25	Bush	34	Spring washer – 2 off
						35	Washer – 2 off

9 Rear suspension assembly: general information

1   On the LC models the conventional pivoted rear fork, or 'swinging arm' is replaced by a welded tubular sub-frame which allows a single large suspension unit to control the rear wheel movements. The unit is mounted between the uppermost point of the sub-frame and the frame top tube, running almost horizontally beneath the dual seat. This cantilever suspension arrangement is known by Yamaha as 'Monocross'.

2   The suspension unit is of conventional construction. An oil-filled damper unit is fitted inside the main spring and is designed to provide control of rear wheel movement by the usual expedient of allowing oil to be forced through a small orifice. A quantity of nitrogen is contained in a high pressure chamber at one end of the unit and this allows for oil movement without the risk of aeration of the damping oil and the consequent loss of damping effect. A free-floating piston and O-ring keep the oil and nitrogen separate.

3   The damper incorporates a floating valve arangement which is designed to reduce damping effect when the damper is moved suddenly. This allows the suspension to move rapidly under the forceful input of a large rut for example, and allows a higher degree of suspension compliance than is normally possible. The floating valve is effectively a spring steel disc, much like a thin washer, and is trapped between two seating faces of the damper piston. Its operation is best illustrated by referring to the accompanying figure in which various damping rates are shown.

4   In A, the unit is under fairly gentle compression and the damping oil flows through the four notches in the valve. If a large bump is encountered the damping rate is too high to allow the unit to react quickly so the valve is arranged to deform, as shown in B, to allow a much greater oil flow through the piston. The same system applies under rebound as shown in C, during gentle movement and D, where the valve has deformed to allow rapid movement at the rear wheel.

5   The suspension unit is adjustable for spring preload via a collar which incorporates a cam arrangement. Adjustment is effected by removing the seat and the protective tray and rotating the collar with the aid of a C-spanner. No manual damping adjustment is provided.

10 Rear sub-frame: examination and renovation

1   The rear sub-frame pivots on headed bushes and a hardened steel sleeve in much the same manner as most conventional pivoted rear forks. Before the assembly can be checked for wear it will be necessary to remove the rear wheel (see Chapter 6, Section 10) and then release the suspension unit from the front or rear mounting points. Of the two, the front mounting is more easily accessible. Straighten and remove the small split pin from the end of the mounting bolt. The nut can now be removed and the bolt displaced to free the unit.

2   Grasp the fork ends of the sub-frame unit and try to move it from side to side. Any clearance should be minimal and must not exceed 1.0 mm (0.04 in). Movement in excess of this means that the bushes are in need of attention and that the sub-frame must be removed.

3   Remove the pivot bolt retaining nut, then tap out the pivot bolt using a long screwdriver or a metal rod, taking care not to damage its threads. It will now be possible to pull the subframe partially clear of the machine, but note that the extent of its movement will be impaired if fitted with an endless final drive chain. If it is wished to remove the sub-frame to a workbench it will be necessary to remove the left-hand engine cover so that the chain can be disengaged from the gearbox sprocket.

4   Pull off the dust seals which cover the ends of the bushes, noting the location of any shims found beneath them. These must be refitted in their original locations during reassembly. Displace the inner pivot sleeve from the sub-frame bore to

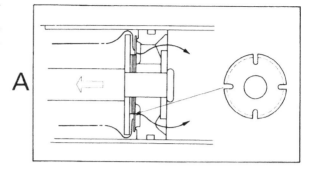

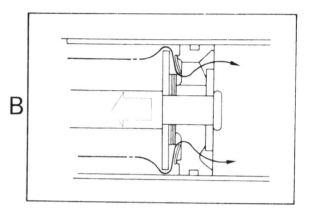

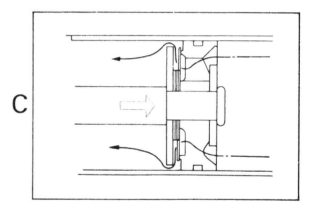

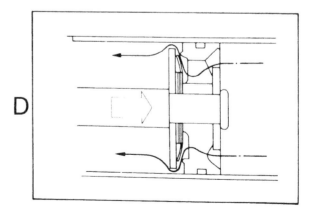

Fig. 5.4 Rear suspension unit operation

expose the bushes. Note that the bushes will inevitably get damaged during removal and must therefore always be renewed if they are removed.

5   The old bushes can be removed by passing a long drift through the cross-tube and driving out the opposing bush. Make sure that the sub-frame is well supported to avoid any risk of distortion during the removal operation. Tap around the bush so that it is driven out squarely. The remaining bush can now be removed by the same method.

6   Yamaha insist that the new bushes must not be driven into place, but must be pushed home using a suitable press. Few owners will have the use of such equipment, but bush fitting is possible by employing a drawbolt arrangement.

7   It will be necessary to obtain a long bolt or a length of threaded rod from a local engineering works or some other supplier. The bolt or rod should be about 1 inch longer than the combined length of the cross tube and one bush. Also required are suitable nuts and two large and robust washers. In the case of threaded rod, fit one nut to one end of the rod and if required, stake it in place for convenience.

8   Fit one of the washers over the bolt or rod so that it rests against the head, then pass the assembly through the cross-tube. Over the projecting end place the bush, which should be greased to ease installation, followed by the remaining washer and nut. Holding the bush to ensure that it is kept square, slowly tighten the nut so that the bush is drawn into the cross-tube. Once it is fully home, remove the drawbolt arrangement and repeat the sequence to fit the remaining bush.

9   Examine the inner pivot sleeve for signs of wear or corrosion. Light corrosion may be removed by careful use of fine abrasive paper, taking great care not to remove any significant amount of metal, particularly at the ends that are supported by the bushes. More serious corrosion or any significant amount of wear will necessitate the renewal of the sleeve.

10 When refitting the sub-frame, grease the bearings and inner sleeve, pushing grease into the space between the bearings and also into the centre of the sleeve. Renew the dust seals if these appear worn or damaged, and remember to refit the shims in their original locations. Note the following torque wrench settings during reassembly.

Sub-frame pivot bolt	6.5 kgf m (46.8 lbf ft)
Suspension unit mountings	4.0 kgf m (28.0 lbf ft)
Rear wheel spindle	4.5 kgf m (32.0 lbf ft)

10.1a Remove split pin ...

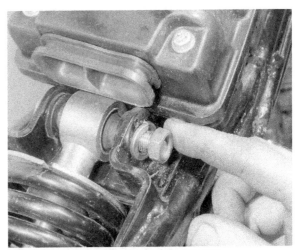

10.1b ... and release bolt to free suspension unit

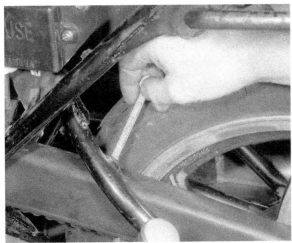

10.3a Release mounting bolt ...

10.3b ... and remove chainguard

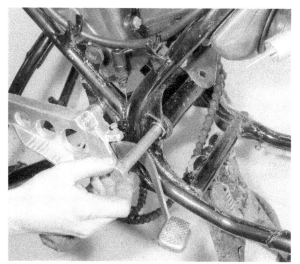

10.3c Remove nut and withdraw the pivot shaft ...

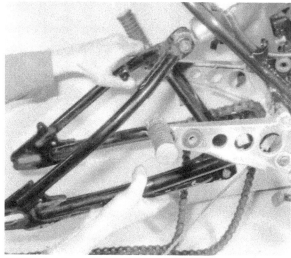

10.3d ... to allow subframe to be removed

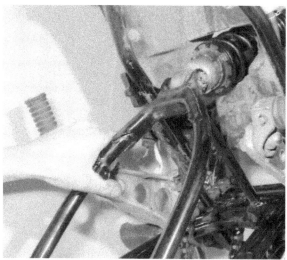

10.3e Suspension unit will pass through gap in mudguard

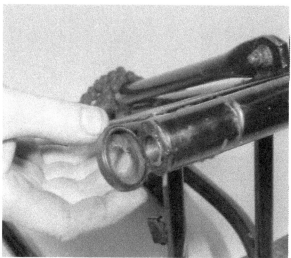

10.4 Remove end covers, noting any shims fitted

10.5 Old bushes can be drifted out of subframe

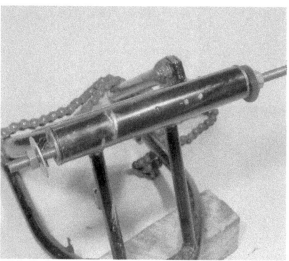

10.8a To fit new bushes, assemble drawbolt as shown ...

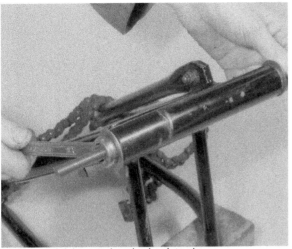

10.8b ... and tighten to draw bushes into place

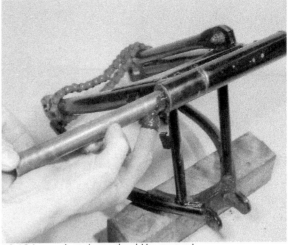

10.10 Inner pivot sleeve should be greased

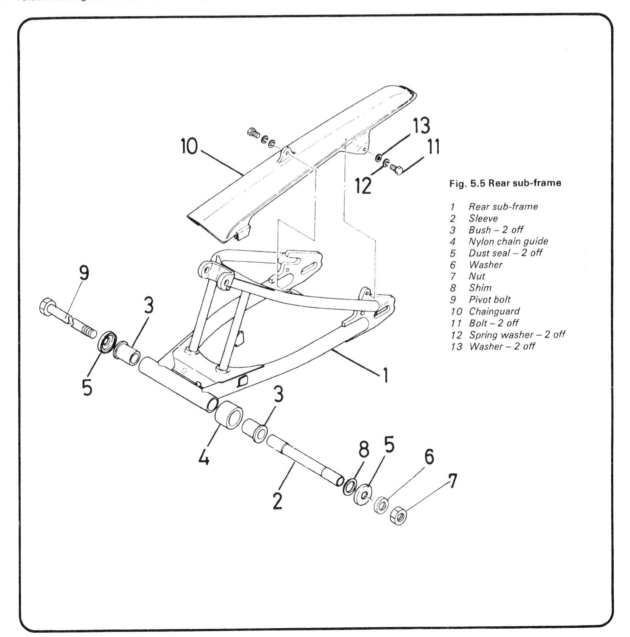

Fig. 5.5 Rear sub-frame

1   Rear sub-frame
2   Sleeve
3   Bush – 2 off
4   Nylon chain guide
5   Dust seal – 2 off
6   Washer
7   Nut
8   Shim
9   Pivot bolt
10 Chainguard
11 Bolt – 2 off
12 Spring washer – 2 off
13 Washer – 2 off

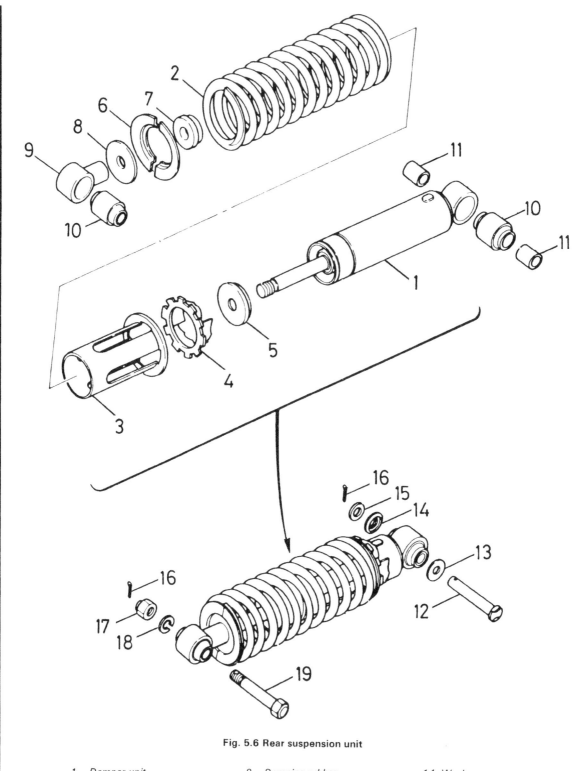

Fig. 5.6 Rear suspension unit

1 Damper unit	8 Damping rubber	14 Washer
2 Spring	9 Lower mounting	15 Washer
3 Spring guide	10 Bush	16 Split pin
4 Adjusting ring	11 Sleeve – 2 off	17 Nut
5 Damping rubber	12 Clevis pin	18 Spring washer
6 Split collet	13 Washer	19 Bolt
7 Bush		

11 Rear suspension unit: examination

1   As mentioned previously, the rear suspension unit is of sealed construction and thus cannot be repaired in the event of failure. Should the damping effect become reduced as a result of wear it is advisable to obtain a new replacement unit well in advance of intended renewal. At the time of writing a waiting time of several weeks is not uncommon for a replacement unit.

2   It should be possible to obtain the bare damper insert, allowing the existing spring and fittings to be re-used. To remove the spring it will be necessary to compress the spring to allow the split collet halves to be displaced. It is recommended that this operation is entrusted to the Yamaha dealer who is supplying the new damper unit since he will have the necessary facilities to complete the job safely. It should be noted that the spring is under quite a lot of tension and no attempt should be made to remove it unless a **safe** method of compressing it is available.

3   Should it become necessary to dispose of the cylinder do not just throw it away. It is first necessary to release the gas pressure and the manufacturers recommend that the following procedure is followed.

4   Refer to the accompanying figure and mark a point 10 – 15 mm above the bottom of the cylinder. Place the unit securely in a vice. Wearing proper eye protection against escaping gas and/or metal particles, drill a 2 – 3 mm hole through the previously marked point on the cylinder.

5   Refitting the suspension unit is the reversal of the removal procedure, noting the following points. Smear a thin layer of grease on the inner faces of the washers and thrust covers. Always use new split-pins on reassembly.

11.1a Subframe suspension mounting bolt is located by raised stop (arrowed)

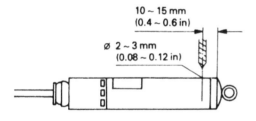

Fig. 5.7 Position of drilling on rear suspension unit

10 ~ 15 mm
(0.4 ~ 0.6 in)

Ø 2 ~ 3 mm
(0.08 ~ 0.12 in)

11.1b Bolt can be withdrawn to free suspension unit

11.2a Suspension unit employs variable rate spring ...

11.2b ... which is retained by split collets

11.5 Mounting bush is protected by end seals

12 Centre stand: examination and maintenance

1   The centre stand is an important but largely neglected feature of most motorcycles. It is important to check the stand for wear or damage from time to time, as failure of the stand can result in costly repair bills. Check that the stand mounting shaft is secure and in good condition, and that it is kept adequately lubricated.
2   Check that the return spring is in good condition. A broken or weak spring may cause the stand to fall whilst the machine is being ridden, and catch in some obstacle, unseating the rider.

13 Prop stand: examination

1   The prop stand is attached to a lug welded to the left-hand lower frame tube. An extension spring anchored to the frame ensures that the stand is retracted when the weight of the machine is taken off the stand.
2   Check that the pivot bolt is secured and that the extension spring is in good condition and not overstretched. An accident is almost certain if the stand extends whilst the machine is on the move.

14 Footrests: examination and renovation

1   The front and rear footrests are mounted on alloy brackets on each side of the machine, these doubling as mounting points for the silencers, and for the gearchange and brake levers. The alloy plates may be removed after the latter parts have been detached, though this should rarely prove necessary.
2   The footrests are of the folding type and are unlikely to require frequent attention. Little can be done to repair them, and in the event of extreme wear or damage, the affected parts should be renewed. If the footrests are damaged in an accident, it is possible to dismantle the assembly into its component parts. Detach each footrest from the mounting plate by withdrawing the split pin and pulling out the clevis pin.

15 Rear brake pedal: examination and renovation

1   The rear brake pedal pivots on a pin projecting from the alloy footrest plate and is secured by a washer and split pin. If damaged, the pedal can be removed after the split pin has been extracted.
2   The pedal can be straightened after heating it to a dull cherry red using a blowtorch or blowlamp. If there is any doubt about the condition of the pedal it should be renewed as a safety precaution.
3   The rear brake pedal is returned to its normal position by an extension spring. This should be checked to ensure that it is not stretched, and pulls the brake off cleanly.

16 Speedometer and tachometer heads: removal and replacement

1   The instrument heads can be removed as a complete assembly as described in Section 3, paragraphs 2 and 3. Alternatively, the two instruments together with the warning lamps can be removed from the moulded base after removing the two small domed nuts which secure each one and releasing the drive cables. As the assembly is lifted upwards, it will be necessary to pull out the warning and illumination lamps, noting the position of each one as a guide during reassembly.
2   Apart from defects in either the drive or drive cables, a speedometer or tachometer which malfunctions is difficult to repair. Fit a replacement or alternatively entrust the repair to a competent instrument repair specialist.
3   Remember that a speedometer in correct working order is a statutory requirement in the UK. Apart from this legal necessity, reference to the odometer readings is the most satisfactory means of keeping pace with the maintained schedules.

17 Speedometer and tachometer drive cables: examination and maintenance

1   It is advisable to detach the drive cable(s) from time to time in order to check whether the outer coverings are damaged or compressed at any point along their run. Jerky or sluggish movements can obtain be traced to a damaged drive cable.
2   It is not possible to effect a satisfactory repair to a damaged or broken drive cable, and in this event the complete cable must be renewed.

18 Speedometer and tachometer drives: location and examination

1   Drive to the speedometer is taken from a small gearbox mounted on the front wheel and anchored to the left-hand fork leg. The gearbox rarely gives rise to problems provided that it is kept well greased whenever the front wheel is removed. In the event of failure, the gearbox must be replaced as a unit, no individual parts are available.
2   The tachometer is driven from a mechanism incorporated in the engine unit, its takeoff point being near the rear of the crankcase. Part of the drive is contained within the crankcase halves and it is therefore necessary to remove and dismantle the engine unit to gain access to it. Further information on the tachometer drive will be found in Chapter 1.

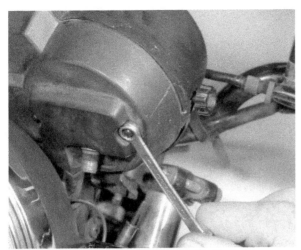

16.1a To remove instrument panel, release nuts ...

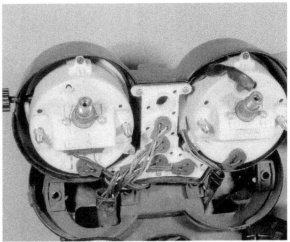

16.1b ... and lift panel away from backplate

Chapter 6 Wheels, brakes and tyres

Contents

General description 1
Front wheel: examination and renovation 2
Disc brake: pad renewal 3
Front disc brake: removing, overhauling and refitting the
caliper units 4
Front disc brake: master cylinder examination and
renovation 5
Front disc brake: hydraulic hose examination 6
Brake disc: removal, examination and replacement 7
Bleeding the hydraulic brake system 8
Front wheel: removal and replacement 9
Rear wheel: removal and replacement 10
Front wheel bearings: examination and renovation 11
Rear wheel bearings: examination and renovation 12

Cush drive assembly: examination and renovation –
RD350LC 13
Cush drive assembly: examination and renovation –
RD250LC 14
Rear wheel sprocket: removal, examination and
replacement – RD350LC 15
Rear wheel sprocket: removal, examination and
replacement – RD250LC 16
Rear brake: examination and renovation 17
Final drive chain: examination and lubrication 18
Tyres: removal, repair and refitting 19
Valve cores and caps 20
Wheel balancing 21

Specifications

Wheels

	Front	Rear
Type	Cast alloy	Cast alloy
Size	1.85 x 18	1.85 x 18
Maximum runout at rim:		
Axial	2.0 mm (0.080 in)	2.0 mm (0.080 in)
Radial	2.0 mm (0.080 in)	2.0 mm (0.080 in)

Tyres

Size	3.00S-18-4PR	3.50S-18-4PR

Tyre pressures

Normal riding	1.75 kg/cm^2 (25 psi)	2.00 kg/cm^2 (28 psi)
High speed	2.00 kg/cm^2 (28 psi)	2.25 kg/cm^2 (32 psi)

Brakes

Type:		
RD250 LC	Single hydraulic disc	sls drum
RD350 LC	Twin hydraulic disc	sls drum
Disc diameter	267 mm (10.5 in)	–
Disc thickness	5 mm (0.19 in)	–
Service limit	4.5 mm (0.18 in)	–
Pad thickness	6.8 mm (0.27 in)	–
Service limit	0.8 mm (0.03 in)	–
Caliper bore diameter	38.18 mm (1.5 in)	–
Master cylinder bore diameter:		
RD250 LC	12.70 mm (0.5 in)	–
RD350 LC	15.87 mm (0.62 in)	–
Hydraulic fluid	DOT 3/SAE J1703	–

Rear drum brake

Drum diameter	–	180 mm (7.09 in)
Lining thickness	–	4 mm (0.16 in)
Service limit	–	2 mm (0.08 in)
Spring free length	–	68 mm (2.68 in)

1 General description

The Yamaha LC models are equipped with cast aluminium alloy wheels carrying conventional tubed tyres. The 250cc version is fitted with a single hydraulic disc brake, the 350 cc model having twin discs. The rear brake on both machines is a single leading shoe (sls) drum brake. The rear wheel differs somewhat between the two models, but only with respect to the cush drive and sprocket mounting arrangements.

2 Front wheel: examination and renovation

1   Carefully check the complete wheel for cracks and chipping, particularly at the spoke roots and the edge of the rim. As a general rule a damaged wheel must be renewed as cracks will cause stress points which may lead to sudden failure under heavy load. Small nicks may be radiused carefully with a fine file and emery paper (No 600 – No 1000) to relieve the stress. If there is any doubt as to the condition of a wheel, advice should be sought from a reputable dealer or specialist repairer.
2   Each wheel is covered with a coating of lacquer, to prevent corrosion. If damage occurs to the wheel and the lacquer finish is penetrated, the bared aluminium alloy will soon start to corrode. A whitish grey oxide will form over the damaged area, which in itself is a protective coating. This deposit however, should be removed carefully as soon as possible and a new protective coating of lacquer applied.
3   Check the lateral and radial run out at the rim by spinning the wheel and placing a fixed pointer close to the rim edge. If the maximum run out is greater than 2.0 mm (0.080 in) the manufacturer recommends that the wheel be renewed. This is, however, a counsel of perfection; a run out somewhat greater than this can probably be accommodated without noticeable

effect on steering. No means is available for straightening a warped wheel without resorting to the expense of having the wheel skimmed on all faces. If warpage was caused by impact during an accident, the safest measure is to renew the wheel complete. Worn wheel bearings may cause rim run out. These should be renewed.

3 Disc brake: pad renewal

1   The caliper and pad arrangement on both models is similar, with the obvious exception that the 350 cc model employs a twin disc arrangement and this has two calipers and two sets of pads to deal with. It should be noted at this stage that the calipers and pads have been modified during the production run. In the case of RD 250 LC models up to engine number 4L1-0082231, and RD 350 LC models up to 4LO-007721 it is essential to use brake pad kit 4LO-W0045-00. All subsequent models require brake pad kit 4W1-W0045-00. It is important to ensure that the correct pads are fitted to each type of caliper. If pads other than Yamaha originals are fitted it is worthwhile noting that EBC Ltd produce replacements for both types, these being numbered FA64 for the original type and FA70 for the later type.
2   To remove the brake pads, release the spring clip which secures the pad retaining pin. The clip is located on the inboard end of the pin and may differ in design from the type shown in the accompanying photographs. It will also be noted that the caliper has been removed for clarity; in practice the pads can be renewed with the caliper and wheel in position.
3   Once the clip has been removed the retaining pin can be displaced and the pads removed. If either pad is worn to 0.8 mm (0.03 in) or less, the pads should be renewed as a pair. Yamaha recommend that the spring clip, retaining pin and shims are renewed every time the pads are renewed.

3.2 Design of security clip may vary between models

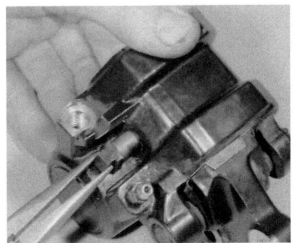

3.3a Remove clip and withdraw pad support pin

3.3b Pads may now be lifted away from caliper

3.3c Note pad backing shim ...

3.3d ... and anti-rattle shim in caliper body

3.3e Centre groove indicates maximum wear depth

4 Front disc brake: removing, overhauling and refitting the caliper unit(s)

1   This Section describes the overhaul procedure for the caliper unit fitted to the 250 cc model. The procedure is identical for the 350 cc model except that there are two calipers to be dealt with. In this case, overhaul and refit one before attention is turned to the other to preclude parts from being interchanged.

2   The caliper unit is of the single piston floating type, the caliper body being free to move sideways along a support pin in relation to the caliper mounting bracket. When the handlebar lever is squeezed the piston is displaced pushing the moving pad against the disc. This then causes the caliper body to move slightly in the opposite direction until the fixed pad exerts equal pressure on the opposite face of the disc.

3   If the caliper unit warrants removal for inspection or renovation, it is first necessary to remove and drain the hydraulic hose. Disconnect the union at the caliper. Have a suitable container in which to catch the fluid. At this stage, it is as well to stop the flow of fluid from the reservoir, by holding the front brake lever in against the handlebar. This is easily done using a stout elastic band, or alternatively, a section cut from an old inner tube.

4   Note: Brake fluid will discolour or remove paint if contact is allowed. Avoid this where possible and remove accidental splashes immediately. Similarly, avoid contact between the fluid and plastic parts such as instrument lenses, as damage will also be done to these. When all the fluid is drained from the hose, clean the connections carefully and secure the hose end and fittings inside a clean polythene bag, to await reassembly. As with all hydraulic systems, it is most important to keep each component scrupulously clean, and to prevent the ingress of any foreign matter. For this reason, it is as well to prepare a clean area in which to work, before further dismantling. As in any form of component dismantling, ensure that the outside of the caliper is thoroughly cleaned down.

5   The caliper unit is attached to the inside of the right-hand fork leg by two bolts, which, when removed, will allow the unit to be lifted away. If the caliper is being removed with the front wheel in position, it should be lifted clear of the disc. Remove the brake pads as described in the preceding Section, exposing the piston. The piston may be driven out of the caliper body by an air jet – a foot pump if necessary. Remove the piston seal and dust seal from the caliper body. Under no circumstances should any attempt be made to lever or prise the piston out of the caliper. If the compressed air method fails, temporarily reconnect the caliper to the flexible hose, and use the handlebar lever to displace the piston hydraulically. Wrap some rag around

the caliper to catch the inevitable shower of brake fluid.

6   The caliper mounting bracket can be freed from the caliper body by withdrawing the support pin. This is retained by a small split pin which should be withdrawn and discarded; a new split pin should be fitted during reassembly. Remove the bracket and lift off the retainer and shim, followed by the small coil spring which is fitted into the bracket.

7   Clean each part carefully, using only clean hydraulic fluid. On no account use petrol, oil or paraffin as these will cause the seals to degrade and swell. Keep all components dust free.

8   Examine the piston surface for scoring or pitting, any imperfection will necessitate renewal. The seals should be renewed as a matter of course, re-using an old seal is a false economy. Remember that the safety of the machine is very much dependent on seal and piston condition.

9   Reassemble, again ensuring absolute cleanliness, by reversing the dismantling procedure. Use clean hydraulic fluid as lubricant. Replace the caliper unit on the machine and reconnect the hydraulic hose. Remember that the system will need bleeding before use, by following the instructions given in Section 8 of this Chapter.

4.3 Remove brake hose banjo bolt and drain fluid

4.5a Caliper is retained by two bolts

4.5b Displace piston and remove from caliper body

4.6a Straighten and remove split pin ...

4.6b ... and withdraw pin to free caliper bracket

4.6c Bracket assembly can be lifted clear ...

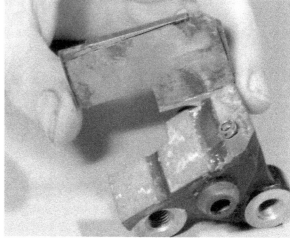

4.6d ... and shim and its retainer removed

4.6e Shim should be removed and both parts cleaned

4.6f Take care not to lose small spring (arrowed)

4.8 Prise out caliper seals and renew them

5 Front disc brake: master cylinder examination and renovation

1   The master cylinder forms a unit with the hydraulic fluid reservoir and front brake lever, and is mounted by a clamp to the right-hand side of the handlebars.

2   The unit must be drained before any dismantling can be undertaken. Place a suitable container below the caliper unit and run a length of plastic tubing from the caliper bleed screw to the container. Unscrew the bleed screw one full turn and proceed to empty the system by squeezing the front brake lever. When all the fluid has been expelled, tighten the bleed screw and remove the tube.

3   Select a suitable clean area in which the various components may be safely laid out, a large piece of white lint-free cloth or white paper being ideal.

4   Remove the locknut and the brake lever pivot bolt to free the lever. As it is lifted away note the small spring which is fitted into the end of the lever blade. Release the front brake switch by pushing a small screwdriver blade into the hole beneath the master cylinder extension which houses it. The switch can now be withdrawn.

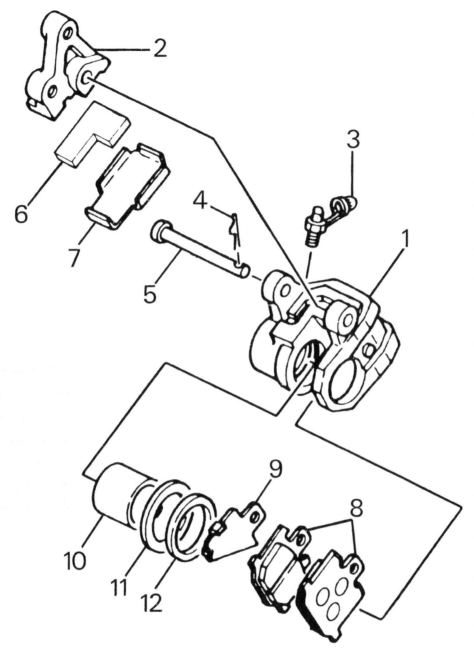

Fig. 6.1 Front brake caliper

1 Caliper	5 Support pin	9 Backing plate
2 Caliper mounting bracket	6 Shim	10 Piston
3 Bleed nipple and cap	7 Shim retainer	11 Piston seal
4 Split pin	8 Brake pads	12 Boot

5 Remove the two bolts which hold the master cylinder clamp half to the body and then lift the master cylinder away. Remove the cover and empty the reservoir. If it is still connected, remove the banjo bolt and free the hydraulic hose from the master cylinder body.

6 Pull off the dust seal from the end of the piston bore to expose the piston end and the circlip which retains it. Remove the circlip to free the piston. If the piston tends to stick in the bore it can be pulled clear using pointed-nose pliers. As the piston is removed the main seal and spring will be released.

7 Examine the piston and seals for scoring or wear and renew if imperfect. Excessive scoring may be due to contaminated fluid, and if this is suspected, it is probably worth checking the condition of the caliper seals and piston. It is recommended that the piston seal is renewed as a matter of course because leakage often occurs once it has been disturbed.

8 Reassemble carefully, using hydraulic fluid as a lubricant on seals and piston, reversing the dismantling sequence. Make sure the rubber boot is fitted correctly, and that the unit is clamped securely to the handlebars. Reconnect the hydraulic

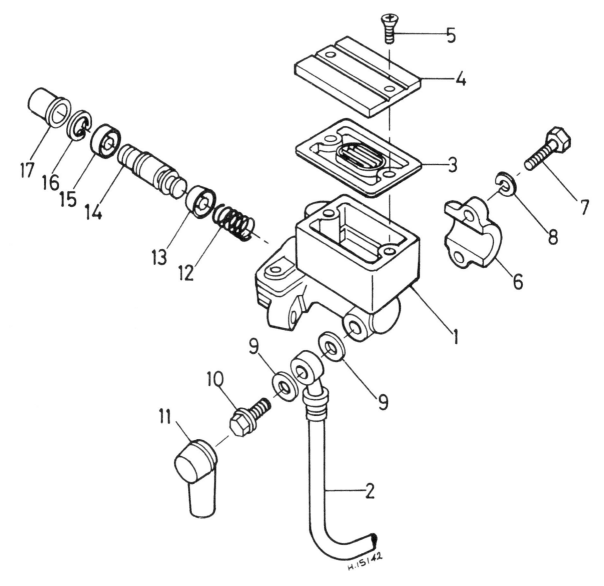

Fig. 6.2 Front brake master cylinder

1	Master cylinder body	5	Screw – 2 off
2	Hydraulic hose	6	Handlebar clamp
3	Diaphragm	7	Bolt – 2 off
4	Cover	8	Spring washer – 2 off

9	Sealing washer – 2 off	13	Main seal
10	Banjo union bolt	14	Piston
11	Rubber cover	15	Primary cup
12	Spring	16	Circlip
		17	Dust seal

hose, tightening the banjo union bolt to the recommended torque setting. Refill the reservoir remembering to top up after the system has been bled by following the procedure given in Section 8 of this Chapter.

6 Front disc brake: hydraulic hose examination

1   An external hose is employed to transmit the hydraulic pressure from the master cylinder to the caliper unit when pressure is applied to the front brake lever.

2   The hose, of the flexible armoured type, must withstand considerable pressure in service, and whilst it is easily ignored, it should be checked carefully as a sudden failure can be potentially fatal. Look not only for signs of chafing against the

wheel or fork leg, but also for any stains due to fluid seepage from cracks in the hose or from the connections at either end.

7 Brake disc: removal, examination and replacement

1   The brake disc, or discs in the case of the 350 cc model, are retained to the hub by six bolts and double tab washers. Should removal prove necessary it will first entail the removal of the front wheel as described in Section 9. Knock back the locking tabs and remove the six bolts. The disc can now be lifted away. When refitting the disc tighten the retaining bolts to 2.0 kgf m (14 lbf ft). Do not omit to secure the bolts by bending over the tab washers.

2   Examination of the disc can be carried out with the wheel

installed. Look for signs of excessive scoring. Some degree of scoring is inevitable, but in severe cases renewal of the disc may prove necessary to restore full braking effect. Check the disc for warpage, which can often result from overheating or impact damage and may cause brake judder. This is best checked using a dial gauge mounted on the fork leg and should not exceed 0.15 mm (0.006 in).

3   The disc thickness should be measured using a vernier caliper or micrometer in several places around the disc surface. The nominal thickness is 5.0 mm (0.19 in) and the disc is in need of renewal if it is worn down to 4.5 mm (0.18 in) or less.

8.4 Bleed tube arrangement is attached as shown

8   Bleeding the hydraulic brake system

1   The method of bleeding a brake system of air and the procedure described below apply equally to either a front brake or rear brake of the hydraulically actuated type.

2   If the brake action becomes spongy, or if any part of the hydraulic system is dismantled (such as when a hose is replaced) it is necessary to bleed the system in order to remove all traces of air. The procedure for bleeding the hydraulic system is best carried out by two people.

3   Check the fluid level in the reservoir and top up with new fluid of the specified type if required. Keep the reservoir at least half full during the bleeding procedure; if the level is allowed to fall too far air will enter the system requiring that the procedure be started again from scratch. Screw the cap onto the reservoir to prevent the ingress of dust or the ejection of a spout of fluid.

4   Remove the dust cap from the caliper bleed nipple and clean the area with a rag. Place a clean glass jar below the caliper and connect a pipe from the bleed nipple to the jar. A clear plastic tube should be used so that air bubbles can be more easily seen. Place some clean hydraulic fluid in the glass jar so that the pipe is immersed below the fluid surface throughout the operation.

5   If parts of the system have been renewed, and thus the system must be filled, open the bleed nipple about one turn and pump the brake lever until fluid starts to issue from the clear tube. Tighten the bleed nipple and then continue the normal bleeding operation as described in the following paragraphs. Keep a close check on the reservoir level whilst the system is being filled.

6   Operate the brake lever as far as it will go and hold it in this position against the fluid pressure. If spongy brake operation has occurred it may be necessary to pump rapidly the brake lever a number of times until pressure is achieved. With pressure applied, loosen the bleed nipple about half a turn. Tighten the nipple as soon as the lever has reached its full travel and then release the lever. Repeat this operation until no more air bubbles are expelled with the fluid into the glass jar. When this condition is reached the air bleeding operation should be complete, resulting in a firm feel to the brake operation. If sponginess is still evident continue the bleeding operation; it may be that an air bubble trapped at the top of the system has yet to work down through the caliper.

7   When all traces of air have been removed from the system, top up the reservoir and refit the diaphragm and cap or cover, as appropriate. Check the entire system for leaks, and check also that the brake system in general is functioning efficiently before using the machine on the road.

8   Brake fluid drained from the system will almost certainly be contaminated, either by foreign matter or more commonly by the absorption of water from the air. All hydraulic fluids are to some degree hygroscopic, that is, they are capable of drawing water from the atmosphere, and thereby degrading their specifications. In view of this, and the relative cheapness of the fluid, old fluid should always be discarded.

9   Great care should be taken not to spill hydraulic fluid on any painted cycle parts; it is a very effective paint stripper. Also, the plastic glasses in the instrument heads, and most other plastic parts, will be damaged by contact with this fluid.

9   Front wheel: removal and replacement

1   Front wheel removal is a fairly straightforward procedure, except in the case of the 350 cc model where the twin calipers tend to impede removal somewhat. It will be noted that it is recommended that the mudguard is removed so that the calipers can be turned outwards to clear the rim and tyre. An alternative method is to remove one of the calipers from the fork leg but on balance the method described here is probably best.

2   Place the machine on its centre stand. Make sure that it is standing securely on a flat surface, then place blocks beneath the front of the engine to raise the front wheel clear of the ground. Unscrew the knurled retaining ring which secures the speedometer drive cable to its gearbox on the front wheel. Pull the cable clear and lodge it clear of the wheel and forks.

3   On the 350 cc model, remove the four mudguard retaining bolts and lift the mudguard clear of the forks.

4   Straighten and remove the split pin which secures the wheel spindle nut. Slacken and remove the nut, then support the wheel while the spindle is displaced and removed. If necessary, tap the spindle through but take care not to damage its threads. Once the spindle has been removed the wheel can be lowered to the ground. In the case of 250 cc machines the wheel can now be lifted clear. On 350 cc machines, turn both brake calipers outwards so that they do not foul the wheel rim and tyre, then remove the wheel.

5   It is good practice to fit a wooden wedge between the brake pads to prevent the caliper pistons from being displaced should the brake lever be operated whilst the calipers are not in position over the disc.

6   The wheel is refitted by reversing the removal sequence. Prior to installation, apply general purpose lithium-based grease such as Castrol LM or equivalent, to the speedometer drive gearbox and the seal faces. Ensure that the speedometer gearbox engages properly with the projection on the fork leg. Tighten the wheel spindle nut to 7.4 kgf m (53.0 lbf ft) and fit a new split pin to secure it.

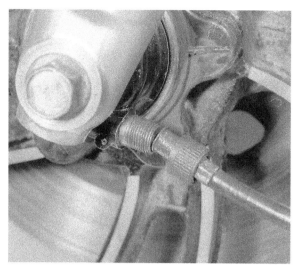

9.2 Release knurled ring and free cable

9.3 Front mudguard must be removed (RD350 LC)

9.4a Remove split pin and wheel spindle nut

9.4b Displace the wheel spindle and remove wheel

9.6a Check that disc engages pads as shown ...

9.6b ... and that speedometer gearbox locates properly

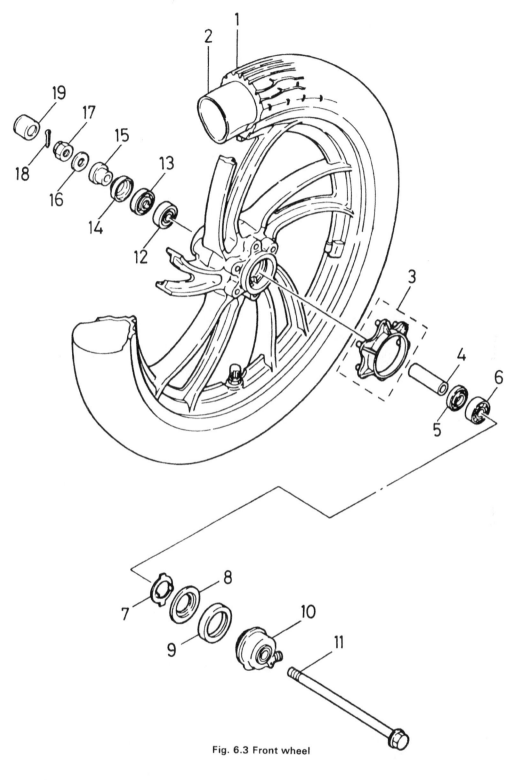

Fig. 6.3 Front wheel

1	Tyre	8	Drive plate retainer	14	Dust cover
2	Inner tube	9	Oil seal	15	Collar
3	Flange – RD250 only	10	Speedometer drive gearbox	16	Washer
4	Centre spacer	11	Wheel spindle	17	Castellated nut
5	Spacer	12	Right-hand bearing	18	Split pin
6	Left-hand bearing	13	Oil seal	19	End cap
7	Speedometer drive plate				

10 Rear wheel: removal and replacement

1   Place the machine securely on its centre stand so that the rear wheel is raised clear of the ground. Unscrew the brake adjuster nut from the brake rod end and displace the rod from the brake arm. Push out the trunnion from the end of the brake arm and fit it and the nut to the end of the brake rod for safe keeping. Detach tbe brake torque arm from the brake backplate lug.

2   Straighten and remove the split pin from the end of the wheel spindle and slacken the wheel spindle nut. Pull the wheel rearwards slightly to allow the adjusters to be pushed downwards through 90°. The wheel can now be pushed forward and the chain disengaged from its sprocket. Withdraw the wheel spindle completely and manoeuvre the wheel clear of the frame noting that the spacer should be displaced if it does not drop clear as the spindle is removed.

3   The wheel is installed by reversing the removal sequence,

having first greased the oil seal lips. Fit the spindle and nut finger-tight, install the chain and check that the chain tension and wheel alignment are correct before final tightening. The chain tension should be checked with the machine *off* its centre stand, the correct amount of free play being 30 - 40 mm (1.2 - 1.6 in) measured at the centre of the lower run. Repeat the check several times with the wheel moved to reposition the chain. There will usually be one point at which the chain is at its tightest, and the chain tension should be set at this point.

4   Adjustment is effected by moving each of the adjuster drawbolts by an equal amount to preserve wheel alignment. As a guide, a row of alignment marks is provided on each side of the wheel spindle slot. Further information on chain adjustment and wheel alignment will be found in Section 18.

5   When adjustment is complete, tighten the wheel spindlle nut to 7.4 kgf m (54 lbf ft). Refit the brake torque arm and tighten its retaining nut to 2.0 kgf m (14 lbf ft). Finally, refit and adjust the brake rod and check the brake lamp switch adjustment. Do not omit to secure the rear wheel spindle nut by fitting a new split pin.

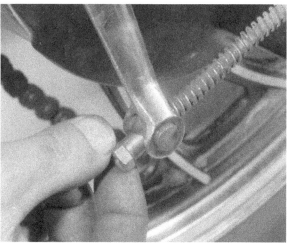

10.1 Remove the rear brake rod adjusting nut

10.2 Free torque arm and remove wheel spindle

10.5a Use new split pin to secure torque arm ...

10.5b ... and rear wheel spindle nuts

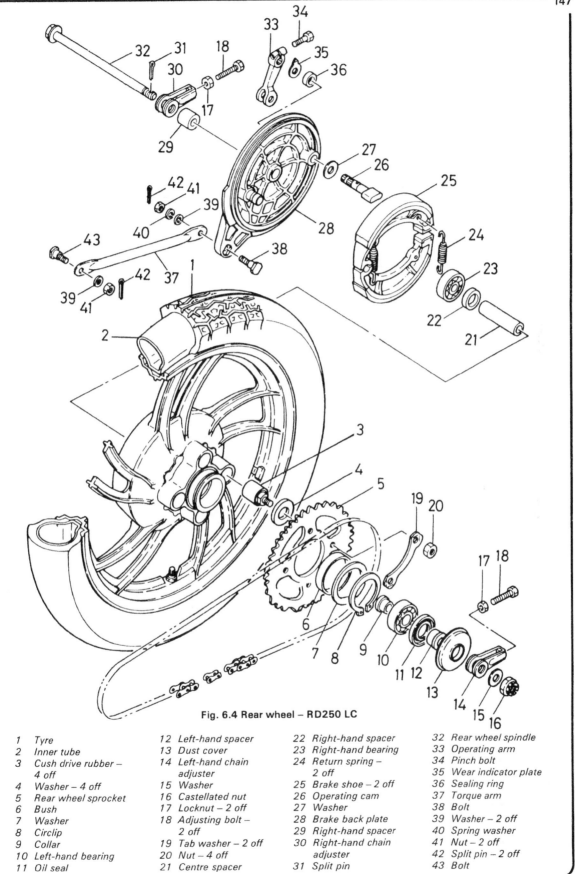

Fig. 6.4 Rear wheel – RD250 LC

1	Tyre	12	Left-hand spacer	22	Right-hand spacer	32	Rear wheel spindle
2	Inner tube	13	Dust cover	23	Right-hand bearing	33	Operating arm
3	Cush drive rubber – 4 off	14	Left-hand chain adjuster	24	Return spring – 2 off	34	Pinch bolt
4	Washer – 4 off	15	Washer	25	Brake shoe – 2 off	35	Wear indicator plate
5	Rear wheel sprocket	16	Castellated nut	26	Operating cam	36	Sealing ring
6	Bush	17	Locknut – 2 off	27	Washer	37	Torque arm
7	Washer	18	Adjusting bolt – 2 off	28	Brake back plate	38	Bolt
8	Circlip	19	Tab washer – 2 off	29	Right-hand spacer	39	Washer – 2 off
9	Collar	20	Nut – 4 off	30	Right-hand chain adjuster	40	Spring washer
10	Left-hand bearing	21	Centre spacer	31	Split pin	41	Nut – 2 off
11	Oil seal					42	Split pin – 2 off
						43	Bolt

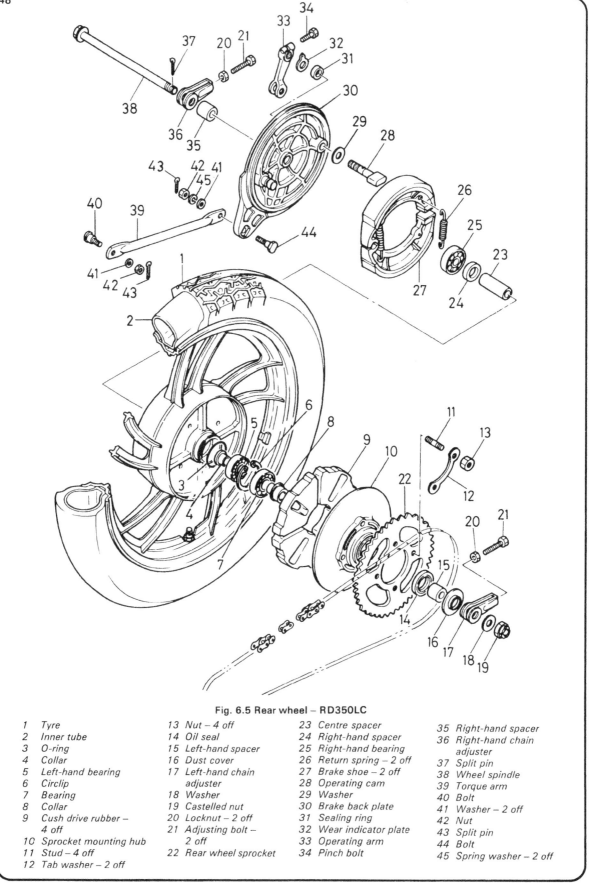

Fig. 6.5 Rear wheel – RD350LC

1	Tyre	13	Nut – 4 off	23	Centre spacer	35 Right-hand spacer
2	Inner tube	14	Oil seal	24	Right-hand spacer	36 Right-hand chain
3	O-ring	15	Left-hand spacer	25	Right-hand bearing	adjuster
4	Collar	16	Dust cover	26	Return spring – 2 off	37 Split pin
5	Left-hand bearing	17	Left-hand chain	27	Brake shoe – 2 off	38 Wheel spindle
6	Circlip		adjuster	28	Operating cam	39 Torque arm
7	Bearing	18	Washer	29	Washer	40 Bolt
8	Collar	19	Castelled nut	30	Brake back plate	41 Washer – 2 off
9	Cush drive rubber –	20	Locknut – 2 off	31	Sealing ring	42 Nut
	4 off	21	Adjusting bolt –	32	Wear indicator plate	43 Split pin
10	Sprocket mounting hub		2 off	33	Operating arm	44 Bolt
11	Stud – 4 off	22	Rear wheel sprocket	34	Pinch bolt	45 Spring washer – 2 off
12	Tab washer – 2 off					

11 Front wheel bearings: examination and renovation

1 The front wheel bearings are an interference fit in the wheel hub, and can be removed by passing a long drift through the centre of one bearing and driving the remaining bearing out from the opposite side. It is advisable to support the wheel on wooden blocks to avoid damage to the disc, or to remove the disc from the wheel.

2 With the wheel suitably supported, pass the drift into position, displacing the spacer between the bearings so that the drift can bear on the inner race of the right-hand bearing. Drive the bearing out of the hub, and remove the spacer.

3 Invert the wheel and drive out the left-hand bearing by inserting a drift of the appropriate size, through the hub. During the removal of either bearing it may be necessary to support the wheel across an open-ended box so that there is sufficient clearance for the bearing to be displaced completely from the hub.

4 Remove all the old grease from the hub and bearings, giving the latter a final wash in petrol. Check the bearings for signs of play or roughness when they are turned. If there is any doubt about the condition of a bearing, it should be renewed.

5 Before replacing the bearings, first pack the hub with new grease. Then drive the bearings back into position, not forgetting the distance piece that separates them. Take great care to ensure that the bearings enter the housings perfectly squarely otherwise the housing surface may be broached.

12 Rear wheel bearings: examination and renovation

1 The rear wheel bearings can be dealt with in much the same manner as described in the preceding Section (see photographs). It will be noted that the rear wheels of the 250 and 350 models differ somewhat. In the case of the 250 cc machines the procedure is very similar to that applied to the front wheel, the wheel being supported on two bearings arranged in a similar fashion in the hub.

2 The 350 cc model has a different cush drive arrangement and employs three wheel bearings. The difference between the two is best illustrated by referring to the accompanying line drawings. The two main bearings are dealt with as described above, the third bearing being accessible after the cush drive hub has been lifted away from the wheel.

11.1a Remove speedometer drive gearbox ...

11.1b ... and prise out grease seal

11.1c Lift away retainer ring ...

11.1d ... and speedometer drive dog

11.1e Moving to RH side, remove headed spacer ...

11.1f ... and seal to expose bearing

11.5a Pack bearing with grease ...

11.5b ... and drive into hub as shown

11.5c Fit distance piece between bearings ...

11.5d ... grease and install remaining bearing

Tyre changing sequence - tubed tyres

 Deflate tyre. After pushing tyre beads away from rim flanges push tyre bead into well of rim at point opposite valve. Insert tyre lever adjacent to valve and work bead over edge of rim.

Use two levers to work bead over edge of rim. Note use of rim protectors

 Remove inner tube from tyre

When first bead is clear, remove tyre as shown

 When fitting, partially inflate inner tube and insert in tyre

Work first bead over rim and feed valve through hole in rim. Partially screw on retaining nut to hold valve in place.

 Check that inner tube is positioned correctly and work second bead over rim using tyre levers. Start at a point opposite valve.

Work final area of bead over rim whilst pushing valve inwards to ensure that inner tube is not trapped

12.1a Remove flanged spacer ...

12.1b ... and seal from hub

12.1c Fit bearing with sealed face outwards ...

12.1d ... and drive it into hub

12.1e Place distance piece into hub bore

12.1f Fit short spacer into bearing as shown ...

12.1g ... then install bearing against hub shoulder

12.1h Grease bearing before fitting seal

13 Cush drive assembly: examination and renovation – RD350 LC

1   The cush drive assembly is contained within the left-hand side of the rear wheel hub. It comprises a set of synthetic rubber buffers, housed within a series of vanes cast in the hub shell. A plate attached to the centre of the rear wheel sprocket has four cast-in dogs which engage with slots in these rubbers, when the wheel is replaced in the frame. The drive to the rear wheel is transmitted via these rubbers, which cushion any surges of roughness in the drive which would otherwise convey the impression of harshness.

2   Examine the rubbers periodically for signs of damage or general deterioration. Renew and fit the rubbers as a set if there is any doubt about their condition; there is no difficulty in removing or replacing them as they are not under compression when the drive plate is attached.

14.2a Remove large circlip ...

14 Cush drive assembly: examination and renovation – RD250 LC

1   The cush drive assembly consists of four tubular rubber bushes located in the hub. The four special pegs retained by nuts on the sprocket locate with these bushes, to give a cushioning effect to the drive sprocket and drive.

2   To obtain access to the bushes, the sprocket has to be removed by detaching its circlip and the four retaining nuts, and then pulling it from the wheel hub. Renewal of the bushes is required when there is excessive free play at the sprocket, often accompanied by snatch in the final drive. It should be noted that due to the effects of rain and road salt the steel pegs tend to corrode and may seize after a considerable length of time. If this occurs, the sprocket complete with pegs should be drawn from the hub using a sprocket puller. A blanking plate or bar, fabricated from mild steel, will have to be made and placed over the bearing. The sprocket puller screw can then bear against the bar.

3   Removal of the flexible bushes is almost impossible without the use of a special expanding extractor. It is recommended that the wheel be returned to a Yamaha dealer who can carry out the work without risk of damage to the wheel.

14.2b ... spacer and ...

14.2c ... the four sprocket retaining nuts

14.2d Sprocket can now be removed ...

14.2e ... followed by the small damper rings

15 Rear wheel sprocket: removal, examination and replacement – RD350 LC

1   The rear wheel sprocket assembly can be removed as a separate unit after the rear wheel has been detached from the frame as described in Section 10 of this Chapter.

2   Check the condition of the sprocket teeth. If they are hooked, chipped or badly worn, the sprocket must be renewed. It is secured to the cush drive plate by four bolts and lock washers.

3   It is considered bad practice to renew one sprocket on its own. The final drive sprockets should always be renewed as a pair and a new chain fitted, otherwise rapid wear will necessitate even earlier renewal on the next occasion.

4   An additional bearing is located within the cush drive plate, which supports the collar through which the rear wheel spindle fits. In common with the wheel bearings, this bearing is a journal ball and when wear occurs, the sprocket will give the appearance of being loose on its mounting bolts. The bearing is a push fit in the cush drive hub and is secured on the inside by a circlip.

5   Remove the circlip and bearing and wash out the latter to remove all traces of the old grease. If the bearing has any play or runs roughly, it must be renewed.

6   If the bearing has not been renewed it should be repacked with grease and refitted in its housing, followed by the circlip. Replace the rear wheel assembly by reversing whichever method was adopted for its removal.

16 Rear wheel sprocket: removal, examination and replacement – RD250 LC

1   The remarks regarding sprocket condition in Section 15 can be applied in general to the 250 cc model. It will be noted, however, that the sprocket is mounted in a somewhat different manner, being retained by four nuts to the rubber-mounted cush drive pegs, and by a large circlip and spacer. The sprocket takes the place of the cush drive hub on the larger models, and reference should be made to Section 14 where some possible complications are discussed.

17 Rear brake: examination and renovation

1   The rear brake is of the single leading shoe (sls) drum type, and is fitted with a brake wear indicator which provides a means of checking the condition of the linings. When the moving pointer reaches the index mark on the back plate, with the brake applied, the linings are in need of renewal. Access to the brake assembly is gained after the rear wheel has been removed as described in Section 10 of this Chapter.

2   When the wheel has been removed, the brake backplate assembly can be lifted away. Remove any accumulated dust from the brake drum by wiping it with a petrol-soaked rag. **Do not** blow the drum out with compressed air. The dust contains asbestos particles which are harmful if inhaled.

3   Measure the lining thickness at its thinnest point and renew the shoes if they are at or near the service limit shown in the Specifications Section. The shoes should also be renewed if they have become contaminated with oil or grease. The lining material is bonded to the shoes and this means that the shoes should be renewed complete. It is not practicable to re-line the shoes at home. When purchasing replacement shoes it is preferable to use genuine Yamaha parts. Many pattern types are available and whilst some may be of good quality, others may be dangerously sub-standard. Be particularly wary of pattern parts supplied with fake Yamaha packaging.

17.2 Lift brake backplate assembly out of hub

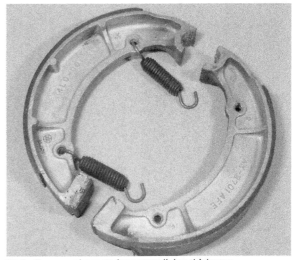

17.3a Remove shoes and measure lining thickness

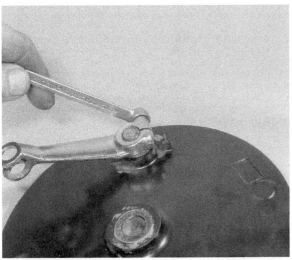

17.3b Remove the brake operating arm ...

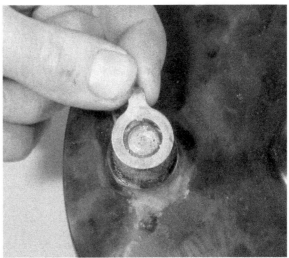

17.3c ... wear indicator pointer ...

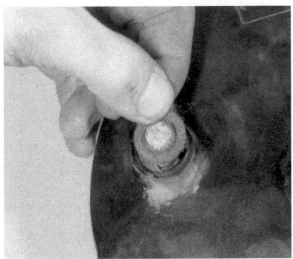

17.3d ... and felt dust seal

17.3e Brake cam can now be removed. Note shim

17.3f Indicator shows wear range of shoes

18 Final drive chain: examination and lubrication

1   The final drive chain is fully exposed, with only a light chainguard over the top run. Periodically the tension will need to be adjusted, to compensate for wear. This is accomplished by placing the machine on the centre stand and slackening the wheel nut on the left-hand side of the rear wheel so that the wheel can be drawn backward by means of the drawbolt adjusters in the fork ends. The rear brake torque arm bolt must also be slackened during this operation.
2   The chain is in correct tension if there is approximately 30 - 40 mm (1.2 - 1.6 in) slack in the middle of the lower run. Always check when the chain is at its tightest point as a chain rarely wears evenly during service. Note that the tension should be checked with the machine resting on its wheels, though it will be necessary to place it back on the centre stand to carry out adjustment.
3   Always adjust the drawbolts an equal amount in order to preserve wheel alignment. The fork ends are clearly marked with a series of horizontal lines above the adjusters, to provide a simple, visual check. If desired, wheel alignment can be checked by running a plank of wood parallel to the machine, so that it touches the side of the rear tyre. If wheel alignment is correct, the plank will be equidistant from each side of the front wheel tyre, when tested on both sides of the rear wheel. It will not touch the front wheel tyre because this tyre is of smaller cross section. See accompanying diagram.
4   Do not run the chain overtight to compensate for uneven wear. A tight chain will place undue stress on the gearbox and rear wheel bearings, leading to their early failure. It will also absorb a surprising amount of power.
5   After a period of running, the chain will require lubrication. Lack of oil will greatly accelerate the rate of wear of both the chain and the sprockets and will lead to harsh transmission. The application of engine oil will act as a temporary expedient, but it is preferable to remove the chain and clean it in a paraffin bath before it is immersed in a molten lubricant such as 'Linklife' or 'Chainguard'. These lubricants achieve better penetration of the chain links and rollers and are less likely to be thrown off when the chain is in motion.
6   The LC models should be equipped with a standard type of DID chain with a spring link, or joining link to facilitate chain removal. To remove the chain for cleaning or renewal the clip should be pushed off using a pair of pliers and the side plate removed. The chain can now be slid apart and run off the sprockets. It is a good idea to keep a length of worn out chain in the workshop. This can be fitted to the end of the machine's

chain and used to pull it off, leaving the worn chain around the sprockets. This will greatly facilitate refitting the chain after cleaning.
7   The machine featured in the photographs throughout this manual posed a few problems since it was fitted with an endless chain. On most machines this means that it is necessary to remove the swinging arm to permit chain removal, but in the case of the LC models with their cantilever sub-frames the problem is compounded because the drain runs through the sub-frame loop. This means that the chain and sub-frame are inextricably intertwined. The only way to separate the two is to break the chain using a chain rivet extractor. A joining link can then be used to prevent future complications. It is uncertain whether this situation is likely to be encountered very frequently, since Yamaha make no reference to the use of endless chains.
8   To check whether the chain is due for replacement, lay it lengthwise in a straight line and compress it endwise so that all the play is taken up. Anchor one end and measure the length. Now pull the chain with one end anchored firmly, so that the chain is fully extended by the amount of play in the opposite direction. If there is a difference of more than $\frac{1}{4}$ inch per foot in the two measurements, the chain should be replaced in conjunction with the sprockets. Note that this check should be made after the chain has been washed out, but before any lubricant is applied, otherwise the lubricant may take up some of the play.
9   When replacing the chain, make sure that the spring link is seated correctly, with the closed end facing the direction of travel.
10   Replacement chains are now available in standard metric sizes from Renold Limited, the British chain manufacturer. When ordering a new chain, always quote the size, the number of chain links and the type of machine to which the chain is to be fitted.

18.7 Endless chain is inseparable from endless frame

19 Tyres: removal, repair and refitting

1   At some time or other the need will arise to remove and replace the tyres, either as a result of a puncture or because replacements are necessary to offset wear. To the inexperienced, tyre changing represents a formidable task, yet if a few simple rules are observed and the technique learned, the whole operation is surprisingly simple.
2   To remove the tyre from either wheel, first detach the wheel from the machine. Deflate the tyre by removing the valve core, and when the tyre is fully deflated, push the bead away from the wheel rim on both sides so that the bead enters the centre well

of the rim. Remove the locking ring and push the tyre valve into the tyre itself.

3   Insert a tyre lever close to the valve and lever the edge of the tyre over the outside of the rim. Very little force should be necessary; if resistance is encountered it is probably due to the fact that the tyre beads have not entered the well of the rim, all the way round. If aluminium rims are fitted, damage to the soft alloy by tyre levers can be prevented by the use of plastic rim protectors.

4   Once the tyre has been edged over the wheel rim, it is easy to work round the wheel rim, so that the tyre is completely free from one side. At this stage the inner tube can be removed.

5   Now working from the other side of the wheel, ease the other edge of the tyre over the outside of the wheel rim that is furthest away. Continue to work around the rim until the tyre is completely free from the rim.

6   If a puncture has necessitated the removal of the tyre, reinflate the inner tube and immerse it in a bowl of water to trace the source of the leak. Mark the position of the leak, and deflate the tube. Dry the tube, and clean the area around the puncture with a petrol soaked rag. When the surface has dried, apply rubber solution and allow this to dry before removing the backing from the patch, and applying the patch to the surface.

7   It is best to use a patch of self vulcanizing type, which will form a permanent repair. Note that it may be necessary to remove a protective covering from the top surface of the patch after it has sealed into position. Inner tubes made from a special synthetic rubber may require a special type of patch and adhesive, if a satisfactory bond is to be achieved.

8   Before replacing the tyre, check the inside to make sure that the article that caused the puncture is not still trapped inside the tyre. Check the outside of the tyre, particularly the tread area to make sure nothing is trapped that may cause a further puncture.

9   If the inner tube has been patched on a number of past occasions, or if there is a tear or large hole, it is preferable to discard it and fit a replacement. Sudden deflation may cause an accident, particularly if it occurs with the rear wheel.

10 To replace the tyre, inflate the inner tube for it just to assume a circular shape but only to that amount, and then push the tube into the tyre so that it is enclosed completely. Lay the tyre on the wheel at an angle, and insert the valve through the rim tape and the hole in the wheel rim. Attach the locking ring on the first few threads, sufficient to hold the valve captive in its correct location.

11 Starting at the point furthest from the valve, push the tyre bead over the edge of the wheel rim until it is located in the central well. Continue to work around the tyre in this fashion until the whole of one side of the tyre is on the rim. It may be necessary to use a tyre lever during the final stages.

12 Make sure there is no pull on the tyre valve and again commencing with the area furthest from the valve, ease the other bead of the tyre over the edge of the rim. Finish with the area close to the valve, pushing the valve up into the tyre until the locking ring touches the rim. This will ensure that the inner tube is not trapped when the last section of bead is edged over the rim with a tyre lever.

13 Check that the inner tube is not trapped at any point. Reinflate the inner tube, and check that the tyre is seating coprrectly around the wheel rim. There should be a thin rib moulded around the wall of the tyre on both sides, which should be an equal distance from the wheel rim at all points. If the tyre is unevenly located on the rim, try bouncing the wheel when the tyre is at the recommended pressure. It is probable that one of the beads has not pulled clear of the centre well.

14 Always run the tyres at the recommended pressures and never under or over inflate. The correct pressures are given in the Specifications Section of this Chapter.

15 Tyre replacement is aided by dusting the side walls, particularly in the vicinity of the beads, with a liberal coating of french chalk. Washing-up liquid can also be used to good effect, but this has the disadvantage, where steel rims are used, of causing the inner surface of the wheel rim to rust.

16 Never fit a tyre that has a damaged tread or sidewalls. Apart from legal aspects, there is a very great risk of a blowout, which can have very serious consequences on a two wheeled vehicle.

17 Tyre valves rarely give trouble, but it is always advisable to check whether the valve itself is leaking before removing the tyre. Do not forget to fit the dust cap, which forms an effective extra seal.

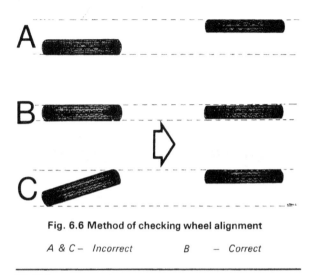

Fig. 6.6 Method of checking wheel alignment

A & C –   Incorrect   B   –   Correct

20 Valve cores and caps

1   Valve cores seldom give trouble, but do not last indefinitely. Dirt under the seating will cause a puzzling 'slow-puncture'. Check that they are not leaking by applying spittle to the end of the valve and watching for air bubbles.

2   A valve cap is a safety device, and should always be fitted. Apart from keeping dirt out of the valve, it provides a second seal in case of valve failure, and may prevent an accident resulting from sudden deflation.

21 Wheel balancing

1   It is customary on all high performance machines to balance the wheels complete with tyre and tube. The out of balance forces which exist are eliminated and the handling of the machine is improved in consequence. A wheel which is badly out of balance produces through the steering a most unpleasant hammering effect at high speeds.

2   Some tyres have a balance mark on the sidewall, usually in the form of a coloured spot. This mark must be in line with the tyre valve, when the tyre is fitted to the inner tube. Even then the wheel may require the addition of balance weights, to offset the weight of the tyre valve itself.

3   If the wheel is raised clear of the ground and is spun, it will probably come to rest with the tyre valve or the heaviest part downward and will always come to rest in the same position. Balance weights must be added to a point diametrically opposite this heavy spot until the wheel will come to rest in ANY position after it is spun.

4   It should be noted that the front wheel must always be checked for balance with the brake disc in position, though it may prove necessary to remove the calipers to eliminate brake drag. The rear wheel can be balanced if required but this is unlikely to have much effect in practice. The balance weights used must be of the correct type for use with Yamaha rims, and to this end should be purchased from a Yamaha dealer. Balance weights are available in 10 gm, 20 gm and 30 gm sizes.

Chapter 7 Electrical system

Contents

General description 1
Testing the electrical system 2
Charging system: checking the output 3
Battery: examination and maintenance 4
Battery: charging procedure 5
Alternator: resistance tests 6
Testing the regulator/rectifier unit 7
Fuses: location and renewal 8
Headlamp: replacing the bulbs and adjusting beam
height 9
Stop and tail lamp: replacement of bulbs 10
Flashing indicator lamps: replacing bulbs 11
Flashing indicator circuit: description and fault diagnosis .. 12

Flasher relay: location and checking 13
Instrument head illumination and warning lamps: bulb
renewal 14
Water temperature gauge: description and testing 15
Temperature gauge sender: testing 16
Horn: location and testing 17
Brake light circuit: testing 18
Brake light switches: location and adjustment 19
Handlebar switches: general 20
Ignition switch: removal and replacement 21
Neutral indicator switch: location and testing 22
Oil level warning lamp: description and testing 23
Wiring: layout and examination 24

Specifications

Battery
Make	Nippon Denso
Model	12N5.5-3B
Voltage	12 volt
Capacity	5.5 Ah

Alternator
Output	9.2 amp or more @ 2000 rpm
	10.5 amp or more @ 5000 rpm

Voltage regulator
Make	Shin Dengen Kougyou
Model	SH 235-12C
Output	14.5 ± 0.5 volts

Rectifier
Make	Shim Dengen Kougyou
Model	SH 235-12C
Type	Full wave
Capacity	15A

Horn
Make	Nikko
Model:	
RD250 LC	SF4-12
RS350 LC	CF-12 (Twin horns fitted)
Amperage (max):	
RD250 LC	3.0A
RD350 LC	2.5A

Flasher relay

Make	Nippon Denso
Model	FU 249CE
Type	Condenser
Wattage	21W x 2 + 3.4W

Fuses

Tail lamp	10A
Flashing indicators	10A
Headlamp	10A
Main	20A

Bulb wattages

Headlamp	60/55W Quartz halogen
Stop/tail	5/21W
Flashing indicators	21W
Parking lamp	3.4W
All instrument/warning lamps	3.4W

1 General description

1 The electrical system is powered by a crankshaft-mounted three-phase alternator located behind the left-hand engine casing. Output from the alternator is fed to a combined rectifier/regulator unit where it is converted from alternating current (ac) to direct current (dc) by the full-wave rectifier section, and the system voltage is regulated to 14.5 ± 0.5 volts by the electronic voltage regulator.

2 The two machines share a common electrical system with the exception of the horn arrangement, the 350 cc model being equipped with twin horns.

2 Testing the electrical system

1 Simple continuity checks, for instance when testing switch units, wiring and connections, can be carried out using a battery and bulb arrangement to provide a test circuit. For most tests described in this Chapter, however, a pocket multimeter should be considered essential. A basic multimeter capable of measuring bolts and ohms can be bought for a very reasonable sum and will provide an invaluable tool. Note that separate volt and ohm meters may be used in place of the multimeter, provided those with the correct operating ranges are available.

2 Care must be taken when performing any electrical test, because some of the electrical components can be damaged if they are incorrectly connected or inadvertently shorted to earth. This is particularly so in the case of electronic components. Instructions regarding meter probe connections are given for each test, and these should be read carefully to preclude accidental damage occurring.

3 Where test equipment is not available, or the owners feels unsure of the procedure described, it is strongly recommended that professional assistance is sought. Errors made through carelessness or lack of experience can so easily lead to damage and the need for expensive replacement parts.

4 A certain amount of preliminary dismantling will be necessary to gain access to the components to be tested. Normally, removal of the seat and side panels will be required, with the possible addition of the fuel tank and headlamp unit to expose the remaining components.

3 Charging system: checking the output

1 In the event that the charging system fails or appears to be over- or under-charging the battery, the system voltage should be checked using a dc voltmeter or a multimeter set on the

0-20 volts dc scale. Remove the side panel to gain access to the battery terminals, noting that the battery leads must **not** be disconnected during the test. **Note**: If the machine is run with the battery disconnected the increased voltage across the alternator terminals will rise, causing damage to the regulator/rectifier unit or to the alternator windings.

2 Connect the positive (red) probe lead to the positive (+) battery terminal and the negative (black) probe lead to the negative (-) battery terminal. Start the engine and note the voltage reading at 2000 rpm. This should be 14.5 ± 0.5 volts if the system is operating correctly. If the voltage is outside this range it will be necessary to check the following, in the order shown below:

a) Battery condition; see Sections 4 and 5
b) Alternator windings; see Section 6
c) Regulator/rectifier; see Section 7

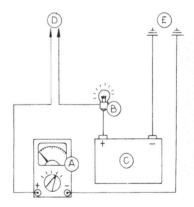

Fig. 7.1 Method of checking the wiring

A	Multimeter	C	Battery
B	Bulb	D	Positive probe
		E	Negative probe

4 Battery: examination and maintenance

1 The battery is housed in a tray located behind the left-hand sidepanel and is retained in position by a bracket which is hinged from the top of the tray and locked in position with a single screw.

2 The transparent plastic case of the battery permits the upper and lower levels of the electrolyte to be observed, without disturbing the battery, by removing the side cover. Maintenance is normally limited to keeping the electrolyte level between the

prescribed upper and lower limits and making sure that the vent tube is not blocked. The lead plates and their separators are also visible through the transparent case, a further guide to the general condition of the battery. If electrolyte level drops rapidly, suspect over-charging and check the system.

3   Unless acid is spilt, as may occur if the machine falls over, the electrolyte should always be topped up with distilled water to restore the correct level. If acid is spilt onto any part of the machine, it should be neutralised with an alkali such as washing soda or baking powder and washed away with plenty of water, otherwise serious corrosion will occur. Top up with sulphuric acid of the correct specific gravity (1.260 to 1.280) only when spillage has occurred. Check that the vent pipe is well clear of the frame or any of the other cycle parts.

4   It is seldom practicable to repair a cracked battery case because the acid present in the joint will prevent the formation of an effective seal. It is always best to renew a cracked battery, especially in view of the corrosion which will be caused if the acid continues to leak.

5   If the machine is not used for a period of time, it is advisable to remove the battery and give it a 'refresher' charge every six weeks or so from a battery charger. The battery will require recharging when the specific gravity falls below 1.260 (at 29°C – 68°F). The hydrometer reading should be taken at the top of the meniscus with the hydrometer vertical. If the battery is left discharged for too long, the plates will sulphate. This is a grey deposit which will appear on the surface of the plates, and will inhibit recharging. If there is sediment on the bottom of the battery case, which touches the plates, the battery needs to be renewed. Prior to charging the battery refer to the following Section for correct charging rate and procedure. If charging from an external source with the battery on the machine, disconnect the leads, or the rectifier will be damaged.

6   Note that when moving or charging the battery, it is essential that the following basic safety precautions are taken:

(a)   Before charging check that the battery vent is clear or, where no vent is fitted, remove the combined vent/filler caps. If this precaution is not taken the gas pressure generated during charging may be sufficient to burst the battery case, with disastrous consequences.

(b)   Never expose a battery on charge to naked flames or sparks. The gas given off by the battery is highly explosive.

(c)   If charging the battery in an enclosed area, ensure that the area is well ventilated.

(d)   Always take great care to protect yourself against accidental spillage of the sulphuric acid contained within the battery. Eyeshields should be worn at all times. If the eyes become contaminated with acid they must be flushed with fresh water immediately and examined by a doctor as soon as possible. Similar attention should be given to a spillage of acid on the skin.

Note also that although, should an emergency arise, it is possible to charge the battery at a more rapid rate than that stated in the following Section, this will shorten the life of the battery and should therefore be avoided if at all possible.

7   Occasionally, check the condition of the battery terminals to ensure that corrosion is not taking place, and that the electrical connections are tight. If corrosion has occurred, it should be cleaned away by scraping with a knife and then using emery cloth to remove the final traces. Remake the electrical connections whilst the joint is still clean, then smear the assembly with petroleum jelly (NOT grease) to prevent recurrence of the corrosion. Badly corroded connections can have a high electrical resistance and may give the impression of complete battery failure.

8   It should be noted that it is almost impossible to test the electrical system with any degree of accuracy unless the battery is in sound condition and fully charged. Many apparent charging system faults can be attributed to an old and worn out battery which can no longer hold a charge. If the battery runs flat when the machine is left unused for a few days it is fairly safe to assume that its useful life has ended. To be certain, have the battery checked under load by an auto-electrical specialist, or check the electrical system using a battery known to be in good condition.

4.2 Battery electrolyte level is visible through battery case

5   Battery: charging procedure

1   The normal charging rate for the 5.5 amp hour battery is 0.5 amps. A more rapid charge, not exceeding 1 amp can be given in an emergency. The higher charge rate should, if possible, be avoided since it will shorten the working life of the battery.

2   Make sure that the battery charger connections are correct, red to positive and black to negative. It is preferable to remove the battery from the machine whilst it is being charged and to remove the vent plug from each cell. When the battery is reconnected to the machine, the black lead must be connected to the negative terminal and the red lead to positive. This is most important, as the machine has a negative earth system. If the terminals are inadvertently reversed, the electrical system will be damaged permanently.

6   Alternator: resistance tests

1   The condition of the alternator stator can be checked by making resistance tests across the stator windings. Unfortunately, the winding resistances are so low that it is beyond the scope of most multimeters to measure them. Also, the resistance difference between the windings is an almost unmeasurable 0.05 ohms. This problem is not helped by the manufacturer's use of identical white output leads which makes it almost impossible to distinguish one connection from its neighbours. For these reasons it must be considered necessary to ask a Yamaha dealer to test the windings.

7   Testing the regulator/rectifier unit

1   The regulator/rectifier unit takes the form of a sealed, finned alloy box mounted immediately inboard of the fuse box. It

should be noted at the outset that, being sealed, there is no possibility of repair should either part of the unit fail.

2 The regulator's operation is tested by performing the charging system output test, as described in Section 3. If the output voltage is incorrect, and the battery, alternator and rectifier are known to be operating correctly, it can be assumed that the unit is faulty and in need of renewal. It is worth having this checked by a Yamaha dealer though.

3 The rectifier is an arrangement of six diodes, connected in a bridge pattern to provide full-wave rectification. This means that the full output of the alternator is converted to dc, rather than half of it, as is the case with simple half-wave rectifiers as used on lightweight machines and mopeds.

4 The condition of the rectifier can be checked using a multimeter, set on its resistance scale, as a continuity tester. Each of the diodes acts as a one-way valve, allowing current to flow in one direction, but blocking it if the polarity is reversed. Perform the resistance check by following the table accompanying Fig. 7.2. If any one test produces the wrong reading the rectifier will have to be renewed.

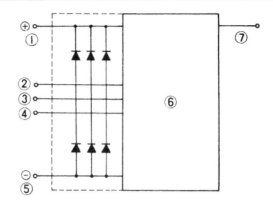

1	Red wire (B)	5	Black wire (E)
2	White wire (U)	6	Regulator/rectifier
3	White wire (V)	7	Brown wire (L)
4	White wire (W)		

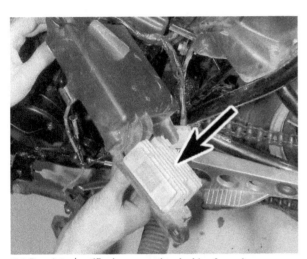

7.1 Regulator/rectifier is mounted on inside of panel

Checking element	Pocket test connecting point		Good	Replace (element shorted)	Replace (element opened)
	(+) (red)	(−) (black)			
D₁	B	U	○	○	×
	U	B	×	○	×
D₂	B	V	○	○	×
	V	B	×	○	×
D₃	B	W	○	○	×
	W	B	×	○	×
D₄	U	E	○	○	×
	E	U	×	○	×
D₅	V	E	○	○	×
	E	V	×	○	×
D₆	W	E	○	○	×
	E	W	×	○	×

○ ..... Continuity
× .... Discontinuity

Fig. 7.2 Testing the rectifier unit

8 Fuses: location and renewal

1 The electrical system is protected by four fuses. A single 20 amp fuse acts as the main protection, with three 10 amp fuses to safeguard the direction indicators, headlamp and tail/stop lamp circuits. The fuses are housed in a moulded plastic box mounted next to the battery and may be reached by unclipping the lid. Spare fuses are clipped to the inside of the lid.

2 If a fuse blows, it should be replaced, after checking to ensure that no obvious short circuit has occurred. If the second fuse blows shortly afterwards, the electrical circuit must be checked thoroughly, to trace the fault.

3 When a fuse blows whilst the machine is running and no spare is available, a 'get you home' remedy is to remove the blown fuse and wrap it in silver paper before replacing it in the fuseholder. The silver plate will restore electrical continuity by bridging the broken fuse wire. This expedient should never be used if there is evidence of a short circuit or other major electrical fault, otherwise more serious damage will be caused. Replace the blown fuse at the earliest possible opportunity, to restore full circuit protection.

8.1 Remove cover to reveal fuses

9 Headlamp: replacing the bulbs and adjusting beam height

1   In order to gain access to the headlamp bulbs it is necessary to first remove the rim, complete with the reflector and headlamp glass. The rim is retained by two screws which pass through the headlamp shell just below the two headlamp mounting bolts.

2   Disconnect the wiring from the bulb terminals by pulling off the connector. The moulded plastic cover can now be peeled off the back of the unit to expose the bulb holder ring. Before the bulb is removed it should be noted that it is of the quartz halogen type. The bulb envelope will become permanently etched if touched, and so it is essential that it is handled by the metal part only. If the quartz envelope is touched accidentally it should be cleaned with methylated spirit and a soft cloth.

3   Remove the retaining ring by twisting it anticlockwise and lift the bulb away. Replacement is a straightforward reversal, the three locating tangs on the bulb flange ensuring that it is correctly aligned.

4   The parking lamp bulb holder can be removed by twisting it anticlockwise. The bulb is a bayonet fitting type rated at 3.4 watts in the UK and 4.0 watts in other countries.

5   The headlamp can be adjusted for both horizontal and vertical alignment. Horizontal adjustment is made via the small screw which passes through the side of the headlamp rim. To set the vertical alignment locate the bolt which passes through the slotted bracket at the rear of the unit. This should be slackened and the headlamp moved to the required position. The bolt is then tightened to retain the setting.

6   In the UK, regulations stipulate that the headlamps must be arranged so that the light will not dazzle a person standing at a distance greater than 25 feet from the lamp, whose eye level is not less than 3 feet 6 inches above that plane. It is easy to approximate this setting by placing the machine 25 feet away from a wall, on a level road, and setting the dipped beam height so that it is concentrated at the same height as the distance of the centre of the headlamp from the ground. The rider must be seated normally during this operation and also the pillion passenger, if one is carried regularly.

10 Stop and tail lamp: replacment of bulbs

1   The combined stop and tail lamp bulb contains two filaments, one for the stop lamp and one for the tail lamp.

2   The offset pin bayonet fixing bulb can be renewed after the plastic lens cover and screws have been removed.

9.1a Headlamp unit is secured by two screws

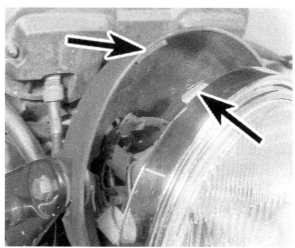

9.1b Note engagement tangs at top of unit

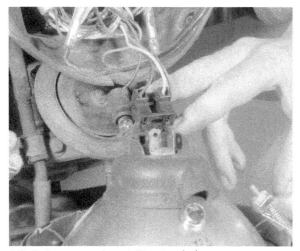

9.2a Bulb connector is push fit on terminals

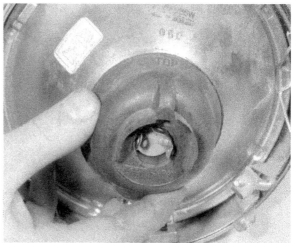

9.2b Pull off rubber shroud from reflector ...

9.3a ... and unscrew bulb retaining ring

9.3b Remove bulb as shown. **Do not** touch the envelope

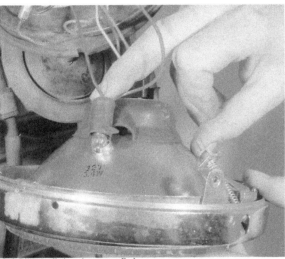

9.4 Pilot bulb has bayonet fitting

9.5a Horizontal adjustment is via screw in rim

9.5b Screw through elongated slot provides vertical adjustment

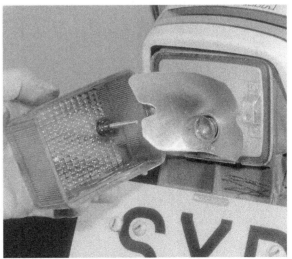

10.1 Rear lamp lens is retained by two screws

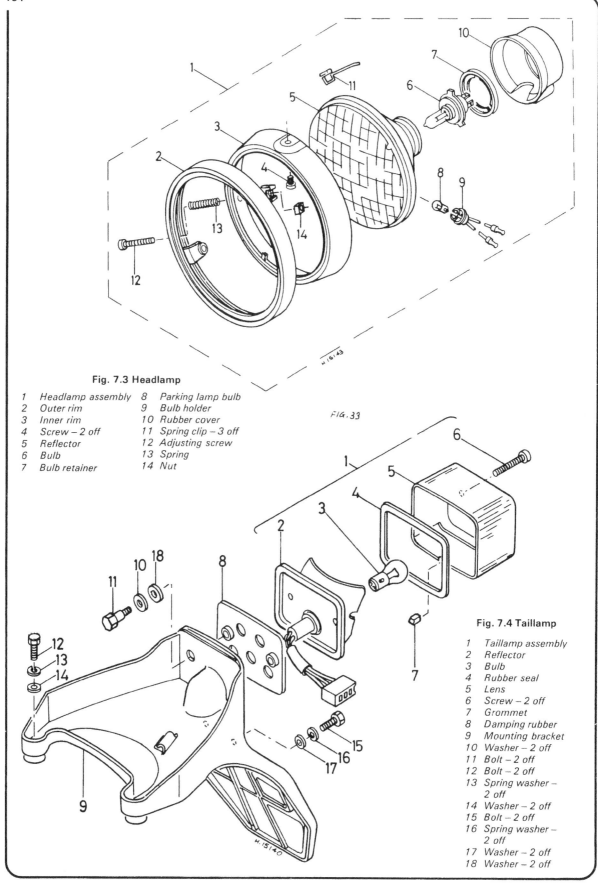

Fig. 7.3 Headlamp

1	Headlamp assembly	8	Parking lamp bulb
2	Outer rim	9	Bulb holder
3	Inner rim	10	Rubber cover
4	Screw – 2 off	11	Spring clip – 3 off
5	Reflector	12	Adjusting screw
6	Bulb	13	Spring
7	Bulb retainer	14	Nut

FIG. 33

Fig. 7.4 Taillamp

1 Taillamp assembly
2 Reflector
3 Bulb
4 Rubber seal
5 Lens
6 Screw – 2 off
7 Grommet
8 Damping rubber
9 Mounting bracket
10 Washer – 2 off
11 Bolt – 2 off
12 Bolt – 2 off
13 Spring washer – 2 off
14 Washer – 2 off
15 Bolt – 2 off
16 Spring washer – 2 off
17 Washer – 2 off
18 Washer – 2 off

11 Flashing indicator lamps: replacing bulbs

1 Flashing indicator lamps are fitted to the front and rear of the machine. They are mounted on short stalks through which the wires pass. Access to each bulb is gained by removing the plastic lens. This is clipped into place and may be carefully eased off using a screwdriver in the slot provided.

12 Flashing indicator circuit: description and fault diagnosis

1 The LC models are equipped with self-cancelling indicators, this function being controlled by a timing circuit and a speedometer sensor which measures the distance travelled. In practice the indicators should switch off after 10 seconds or after 150 metres (164 yards) have been covered. Both systems must switch off before the indicators stop, thus at low speeds the system is controlled by distance, whilst at high speeds, elapsed time is the controlling factor.

2 A speedometer sensor measures the distance covered from the moment that the switch is operated. After the 150 metres have been covered, this part of the system will reset to off. The flasher cancelling unit starts a ten second countdown from the moment that the switch is operated. As soon as both sides of the system are at the off position, the flashers are cancelled. If required, the system may be overriden manually by depressing the switch inwards.

3 In the event of malfunction, refer to the accompanying figure which shows the circuit diagram for the self-cancelling system. The self-cancelling unit is located beneath the fuel tank, immediately to the rear of the steering head. Trace the output leads to the 6-pin connector and disconnect it. If the ignition switch is now turned on and the indicators will operate normally, albeit with manual cancelling, the flasher relay, bulbs, wiring and switch can be considered sound.

4 To check the speedometer sensor, connect a multimeter to the white/green and the black leads of the wiring harness at the 6-pin connector. Set the meter to the ohms x 100 scale. Release the speedometer cable at the wheel end and use the projecting cable end to turn the speedometer. If all is well, the needle will alternate between zero resistance and infinite resistance. If not, the sender or the wiring connections will be at fault.

5 Connect the meter probes between the yellow/red lead and earth, again on the harness side of the 6-pin connector. Check the switch and associated wiring by turning the indicator switch on and off. In the off position, infinite resistance should be shown, with zero resistance in both on positions.

11.1 Indicator lens is clipped to lamp body

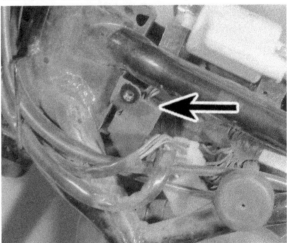

12.3 Indicator cancelling unit (arrowed)

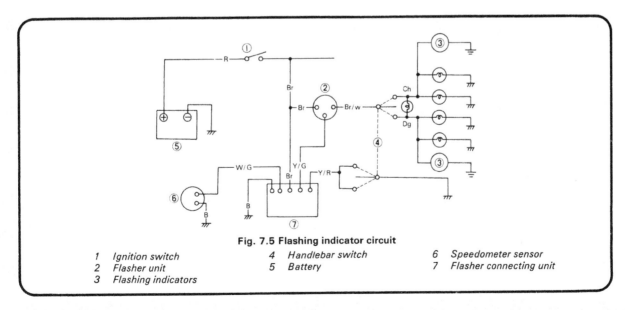

Fig. 7.5 Flashing indicator circuit

1 Ignition switch	4 Handlebar switch	6 Speedometer sensor
2 Flasher unit	5 Battery	7 Flasher connecting unit
3 Flashing indicators		

6   If the above tests reveal no obvious fault in the indicator circuits or the indicators will work only as a manually operated arrangement, the self cancelling system is inoperative and the unit will require renewal.

13 Flasher relay: location and checking

1   If the indicators fails to operate, the nature of the fault gives a good indication of its cause. If the fault is restricted to one set of lamps only and yet the remaining set operate correctly the fault is almost certainly due to a blown bulb, or broken or shorted wiring on that side of the circuit. If the system fails totally, check that this is not caused by the self-cancelling system by unplugging the flasher cancelling unit (see Section 12). If it is found that the latter is at fault it can be left disconnected and the indicators used manually until a replacement unit is obtained.
2   If the fault cannot be attributed to any other cause it will be necessary to renew the flasher relay. It is located on the forward edge of the battery tray and is held in a rubber mounting. The relay is a sealed unit and cannot be repaired if it is faulty. Ensure that the new unit is of the same rating as the standard item. Any variation in its output will affect the flash rate.
3   In the event of a fault occurring in the indicator circuit, the following check list will prove helpful.

Indicators do not work.
   a)   Check bulb
   b)   Right circuit:
        1   Check for 12V on dark green wire to light.
        2   Check for ground on black wire to light assembly.
   c)   Left circuit:
        1   Check for 12V on dark brown wire to light.
        2   Check for ground on black wire to light assembly.
   d)   Right and left circuits do not work:
        1   Check for 12V on brown/white wire to flasher switch on left handlebar.
        2   Check for 12V on brown wire to flasher relay.
        3   Replace flasher relay.
        4   Replace flasher switch.
   e)   Check flasher self cancelling system.
        (Refer to flasher self cancelling system).

14 Instrument head: illumination and warning lamps: bulb renewal

1   Access to the illumination and warning lamp bulbs is gained by releasing the instrument panel from the lower part of its mounting. Start by unscrewing the knurled rings which secure the speedometer and tachometer drive cables to the instrument bases. Remove the four small domed nuts and washers from the underside of the instrument panel. The upper half can now be lifted up to reveal the bulbholders.
2   Remove each bulbholder separately to avoid confusion, or make a sketch to show the wiring colours running to each one. The rubber bulbholders are a push fit in the instrument panel and each one carries a 12 volt 3.4W bulb.

15 Water temperature gauge: description and testing

1   A water temperature gauge is fitted to the LC models to monitor the coolant temperature. The system comprises a gauge built into the tachometer face and a sender unit mounted on the cylinder head. The sender unit is a thermocouple, the resistance of which reduces as it gets hotter. It is connected between the gauge and earth and thus controls the needle position according to engine temperature.
2   In the event that the gauge should fail to operate, check the system in the following sequence:

   a)   Check all wiring connections, cleaning or repairing them as required.
   b)   Check the resistance of the sender unit as described in Section 16, and renew it where necessary.
   c)   If the above tests show the wiring and sender to be in working order it will be necessary to renew the gauge. Unfortunately it is integral with the tachometer head, so it will be necessary to renew the entire assembly. It is suggested that it may be worth trying to obtain a second-hand unit from a breaker; a new instrument will not be cheap.

13.2a Indicator relay is mounted on battery tray

13.2b Connector is push-fit on underside of relay

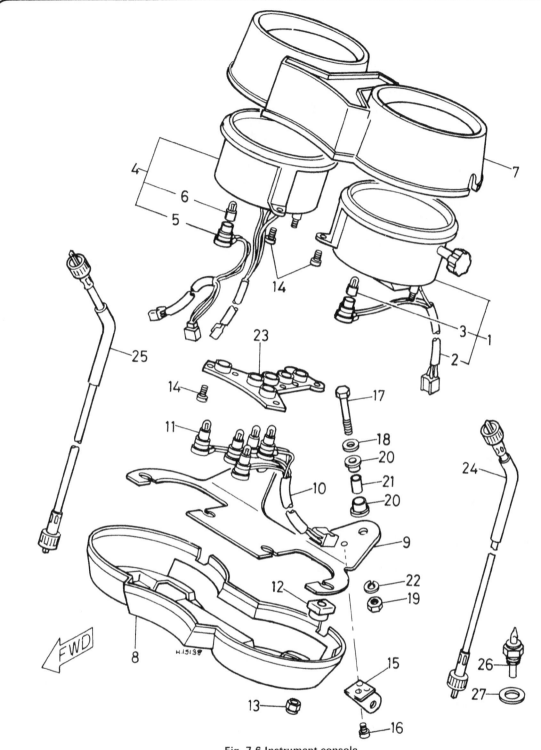

Fig. 7.6 Instrument console

1	Speedometer assembly	10	Pilot lamp wiring	19	Nut – 2 off
2	Speedometer wiring	11	Bulb – 5 off	20	Grommet – 4 off
3	Bulb	12	Damping block – 4 off	21	Spacer – 2 off
4	Tachometer assembly	13	Nut – 4 off	22	Spring washer – 2 off
5	Tachometer bulb wiring	14	Screw – 7 off	23	Pilot lamp mounting
6	Bulb	15	Tachometer bracket	24	Speedometer cable
7	Instrument upper cover	16	Screw	25	Tachometer cable
8	Instrument lower cover	17	Bolt – 2 off	26	Thermo unit
9	Mounting bracket	18	Washer – 2 off	27	Washer

14.1a Remove the four domed nuts ...

14.1b ... and lift the instrument panel away

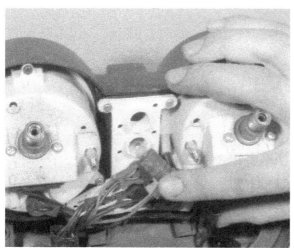

14.1c Bulbs are a push-fit in instrument panel

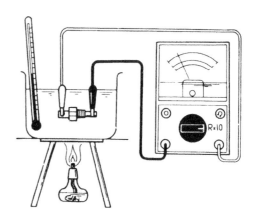

Fig. 7.7 Testing the cooling system temperature gauge sender

16 Temperature gauge sender: testing

1 If the water temperature sender appears to be faulty it can be tested by measuring its resistance at various temperatures. To accomplish this it will be necessary to gather together a heatproof container into which the sender can be placed, a burner of some description (a small gas-powered camping burner would be ideal), a thermometer capable of measuring between 50°C and 120°C (122°F – 248°F) and an ohmmeter or multimeter capable of measuring 0 – 125 ohms with a reasonable degree of accuracy.

2 Fill the container with cold water and arrange the sender unit on some wire so that the probe end is immersed in it. Connect one of the meter leads to the sender body and the other to the terminal. Suspend the thermometer so that the bulb is close to the sender probe.

3 Start to heat the water, and make a note of the resistance reading at the temperature shown in the table below. If the unit does not give readings which approximate quite closely to those shown it must be renewed.

WATER TEMPERATURE	50°C	80°C	100°C	120°C
RESISTANCE	125 Ω	48 ± 4 Ω	28 + 2 Ω	17 Ω

4 Some difficulty may be experienced in attempting to heat the water to 120°C since at atmospheric pressure it boils at 100°C and is reluctant to get much hotter. Without going to extreme lengths to build a pressurised test rig it is best to assume that if the unit works properly between 50°C and 100°C it is likely to work adequately at 120°C.

17 Horn: location and testing

1 The RD250 LC is fitted with a single horn mounted between the fork stanchions on a flexible steel bracket. The RD350 LC differs in that it has twin horns of a slightly different rating. In the event of a fault, follow the test sequence shown

below, using a multimeter set on the appropriate test range.

   a) Test brown horn lead for 12 volts, connecting one meter probe to earth and the other to the lead.

   b) Connect meter probes to pink lead and earth. Test for continuity when the horn button is pressed.

2   If the above tests do not reveal a fault it must be assumed that the horn is faulty. The unit is sealed and cannot be repaired, so a new horn will be required.

18 Brake light circuit: testing

1   In the event of a fault in the brake light circuit follow the test sequence shown below. The switches must be renewed if they are faulty, but occasionally they can be persuaded to work after a good soaking in WD40 or similar.

   a) Check bulb and connections.

   b) Check for 12 volts on yellow lead to brake lamp.

   c) Check for 12 volts on brown lead to front and rear brake switches.

   d) Check black earth lead from lamp unit to frame (continuity test).

19 Brake light switches: location and adjustment

1   All models have a stop lamp switch fitted to operate in conjunction with the rear brake pedal. The switch is located immediately to the rear of the crankcase, on the right-hand side of the machine. It has a threaded body giving a range of adjustment.

2   If the stop lamp is late in operating, slacken the locknuts and turn the body of the lamp in an anticlockwise direction so that the switch rises from the bracket to which it is attached. When the adjustment seems near correct, tighten the locknuts and test.

3   If the lamp operates too early, the locknuts should be slackened and the switch body turned clockwise so that it is lowered in relation to the mounting bracket.

4   As a guide, the light should operate after the brake pedal has been depressed by about 2 cm ($\frac{3}{4}$ inch).

5   A stop lamp switch is also incorporated in the front brake system. The mechanical switch is a push fit in the handlebar lever stock. If the switch malfunctions, repair is impracticable. The switch should be renewed.

20 Handlebar switches: general

1   Generally speaking, the switches give little trouble, but if necessary they can be dismantled by separating the halves which form a split clamp around the handlebars. Note that the machine cannot be started until the ignition cut-out on the right-hand end of the handlebars is turned to the central 'Run' position.

2   Always disconnect the battery before removing any of the switches, to prevent the possibility of a short circuit. Most troubles are caused by dirty contacts, but in the event of the breakage of some internal part, it will be necessary to renew the complete switch.

3   Because the internal components of each switch are very small, and therefore difficult to dismantle and reassemble, it is suggested a special electrical contact cleaner be used to clean corroded contacts. This can be sprayed into each switch, without the need for dismantling.

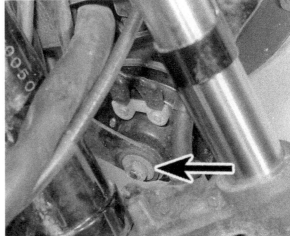

17.1 Horn is mounted between fork legs (RD250 LC model)

19.2 Rear brake light switch location

20.1a Right-hand handlebar switch unit

20.1b Left-hand handlebar switch unit

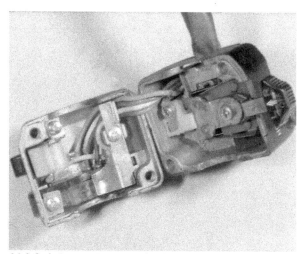

20.3 Switches can be dismantled for cleaning

21 Ignition switch: removal and replacement

1   The combined ignition and lighting master switch is mounted in the warning light panel mounting plate.

2   If the switch proves defective it may be removed after detaching the instrument panel and unscrewing the two mounting screws. Disconnect the two wiring socket connecting the switch to the loom.

3   Reassembly of the switch can be made in the reverse procedure as described for dismantling. Repair is rarely practicable. It is preferable to purchase a new switch unit, which will probably necessitate the use of a different key.

22 Neutral indicator switch: location and testing

1   The neutral indicator lamp is operated by a switch arrangement on the left-hand end of the gear selector drum. Access to the switch is straightforward once the left-hand engine casing has been removed. The switch consists of a triangular plastic cover which is retained by three screws. The inner face of the cover incorporates a circular track into which the fixed contact is set flush. A spring-loaded contact is fitted into the end of the selector drum.

2   If a fault is experienced it is not very likely that the switch will be the cause of it. The following check sequence should be followed:

a)   Check bulb and connections.
b)   Check for 12 volts on sky blue lead at switch terminal.
c)   Check switch continuity. If faulty, clean or renew damaged parts.

23 Oil level warning lamp: description and testing

1   The oil level warning lamp is operated by a float-type switch mounted in the oil tank. The circuit is wired through the neutral switch so that when the ignition is switched on and the machine is in neutral, the lamp comes on as a means of checking its operation. As soon as a gear is selected the lamp should go out unless the oil level is low.

2   In the event of a fault the bulb can be checked by switching the ignition on and selecting neutral. If this proves sound, check for 12 volts on the black/red lead to the switch. If the switch proves to be defective it can be unclipped from the tank and withdrawn.

22.1a Neutral switch cover is retained by three screws

22.1b Remove cover to reveal switch contacts

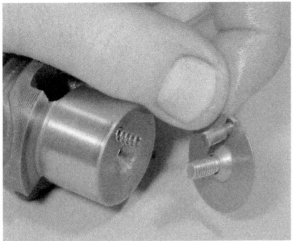

22.1c Plate removed to show contact and spring

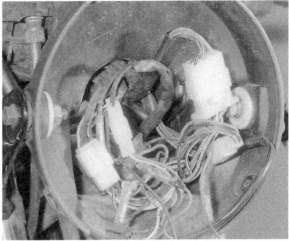

24.1 Wiring is colour coded to aid fault finding

24 Wiring: layout and examination

1 The wiring harness is colour-coded and will correspond with the accompanying wiring diagram. When socket connections are used, they are designed so that reconnection can be made in the correct position only.

2 Visual inspection will usually show whether there are any breaks or frayed outer covering which will give rise to short circuits. Occasionally a wire may become trapped between two components, breaking the inner core but leaving the more resilient outer cover intact. This can give rise to mysterious intermittent or total circuit failure. Another source of trouble may be the snap connectors and sockets, where the connector has not been pushed fully home in the outer housing, or where corrosion has occurred.

3 Intermittent short circuits can often be traced to a chafed wire that passes through or is close to a metal component such as a frame member. Avoid tight bends in the lead or situations where a lead can become trapped between casings.

Wiring diagram component key

1	Front right-hand indicator	20	Temperature gauge sender
2	Front brake stop lamp switch	21	Horn switch
3	Engine kill switch	22	Pass switch
4	Lighting switch	23	Indicator switch
5	Indicator cancelling unit	24	Dip switch
6	Ignition coil	25	Horn
7	CDI unit	26	Front left-hand indicator
8	Spark plugs	27	Ignition switch
9	Neutral indicator switch	28	Parking lamp
10	Alternator	29	Headlamp
11	Oil level sensor	30	Distance sensor
12	Rear right-hand indicator	31	Meter lamp
13	Tail/stop lamp	32	Oil level warning lamp
14	Rear left-hand indicator	33	Neutral indicator lamp
15	Fuse	34	High beam indicator lamp
16	Regulator/rectifier	35	Right-hand indicator warning lamp
17	Rear brake stop lamp switch	36	Left-hand indicator warning lamp
18	Indicator relay	37	Meter lamp
19	Battery	38	Temperature gauge

Wiring diagram – RD250 and 350 LC models

See page 171 for component key

Colour key

B	Black	W	White	
G	Green	Y	Yellow	
L	Blue	Br	Brown	
O	Orange	Ch	Dark brown	
P	Pink	Dg	Dark green	
R	Red	Sb	Light blue	

Index

A

About this manual 2
Acknowledgements 2
Adjustments:-
    autolube pump 26
    brake light 169
    carburettors 25,102
    clutch 25
    engine reassembly 85
    gear selector shaft 73
    headlamp 162
    rear chain 23
    steering head bearings 28, 121
Air cleaner 27, 105
Alternator 42, 79, 160

B

Battery:-
    charging procedure 160
    examination and maintenance 22, 159
    problems 18, 19
Bearings:-
    big-end 58
    main 58
    steering head 28, 126
    wheel:
        front 149
        rear 149
Bleeding the hydraulic brake system 143
Brakes:-
    front disc brake:
        bleeding 143
        disc pad renewal 137
        check 24
        hydraulic hose 142
        master cylinder 140
        refitting caliper units 138
        removing, overhauling 138
    light switches – test and adjustments 169
    problems 17, 18
    rear brake:
        check 24
        examination and renovation 154
Bulbs – replacement:-
    flashing indicators 165
    headlamp 162
    instrument head 166
    stop and tail lights 162

Buying:-
    spare parts 7
    tools 9

C

Cables:-
    clutch 25
    lubrication 23, 25
    oil pump and throttle cable settings 26, 109
    speedometer and tachometer 134
Carburettors :-
    adjustment 25, 102
    dismantling and reassembly 98
    general 97
    removal and refitting 97
    settings 103
    specifications 95
Castrol lubricants 32
Centre stand 134
Chain:-
    cleaning and lubricating 29
    final drive 12, 156
Chapter contents:-
    1   Engine, clutch and gearbox 33
    2   Cooling system 87
    3   Fuel system and lubrication 95
    4   Ignition system 112
    5   Frame and forks 118
    6   Wheels, brakes and tyres 136
    7   Electrical system 158
Checks:-
    charging system 159
    clutch adjustment 25
    coolant level 22
    flasher relay 166
    front and rear brakes 24
    ignition timing 30, 117
    legal 24
    oil pump and throttle cable settings 26, 109
    pre-operation 21
    safety 23
    spark plugs – gap settings 117
    steering and suspension 28
    transmission oil 25
    tyre pressures 22
    wheel and tyre conditions 31
    wiring, connections and switches 114
Cleaning:-
    air filter 27, 105
    chain 29
    exhaust system 28, 103

Clutch:-
    adjustment 25
    drag 14
    examination and renovation 63
    refitting 74
    removal 43
    slip 14
    specifications 34
Coils:-
    exciter coils – testing 116
    ignition coil – testing 116
    pulser coil – testing 116
Cooling system:-
    draining 30, 88
    filling 89
    flushing 89
    general 30, 87
    hoses and connections 91
    radiator and cap 89
    water pump 92
    water temperature gauge 92, 166, 168
Crankcase halves:-
    examination 47
    joining 69
    reassembly 73
    separating 47
Crankshaft:-
    dismantling 47
    reassembly 68
    removal 45
Cush drive:-
    assembly, examination and renovation – RD350 LC 153
    assembly, examination and renovation – RD250 LC 153
Cylinder barrels:-
    examination and renovation 58
    reassembly 79
    removal 40
Cylinder head:-
    examination and renovation 59
    reassembly 79
    removal 40

D

Decarbonising 29
Description – general:-
    cooling system 87
    electrical system 159
    engine, clutch and gearbox 35
    frame and forks 118
    fuel system 96
    ignition system 112
    lubrication system 96
    wheels, brakes and tyres 137
Dimensions and weights 6

E

Electrical system;-
    alternator 42, 79, 160
    battery charging procedure 160
    electrical components 30
    electronic ignition system 114
    fault diagnosis 18, 19
    flasher relay 166
    flashing indicators 165
    fuse location 161
    headlamp 162
    horn 169
    instrument head 166
    lamps 162-165
    lights and electrical components 30
    regulator/rectifier unit 160
    specifications 158
    switches – testing 169, 170
    wiring diagram 172
Engine:-
    bearings:
        big-end 58
        front wheel 149
        main 58
        rear wheel 149
        steering head 126
    clutch:
        adjustment 25
        drag 14
        examination and renovation 63
        refitting 74
        removal 47
        slip 14
        specification 34
    crankcase halves:
        examination 47
        joining 69
        reassembly 73
        separating 47
    crankshaft:
        reassembly 68
        dismantling 47
        removal 45
    cylinder barrels:
        examination and renovation 58
        reassembly 79
        removal 40
    cylinder head:
        examination and renovation 59
        reassembly 79
        removal 40
    decarbonising 29
    dismantling – general 40
    examinaiton and renovation – general 47
    fault diagnosis 11
    lubrication 96
    oil pump – bleeding 109
    oil seals – checking 58, 109
    pistons and rings 59, 79
    primary drive 63, 74
    reassembly – general 63
    specifications 33
    speedometer and tachometer drives 134
    torque wrench settings 34
    valves:
        reed valve induction system 110
        reed valve removal, examination and renovation 110
Exhaust system:-
    cleaning 28, 103
    modifications 103

F

Fault diagnosis:-
    abnormal frame and suspension noise 17
    acceleration 13
    brakes 17
    clutch 14, 15
    electrical 18
    engine 11, 12, 15
    exhaust 15
    gear selection 14
    knocking or pinking 13

overheating 13
poor handling 16
poor running 13
Filter – air 27, 105
Final drive chain 23, 29, 156
Footrests 134
Frame and forks:-
centre stand 134
footrests 134
frame – examination and renovation 127
front forks:
dismantling 123
removal 119
prop stand 134
rear brake pedal 134
rear sub frame 128
rear suspension unit 133
specifications 118
speedometer and tachometer drive cables 134
steering head assembly 28, 121
steering lock 126
Front wheel 137
Fuel system:-
air cleaner 27, 105
carburettor:
adjustment 25, 102
dismantling 98
general 97
removal 97
settings 103
specifications 95
fault diagnosis 11, 12, 13, 16
petrol feed pipes 97
petrol tank 96
petrol tap 96
Fuse location 161

G

Gearbox:-
components – examination and renovation 60
examination and renovation – general 47
fault diagnosis 14, 15
gear input and output shafts 50
specifications 34
Generator – alternator 42, 79, 160
Greasing – general 31

H

Handlebar switches 169
**Headlamp – replacing bulbs and adjusting
beam height** 162
Horn location 169
Hoses – examination and renovation 91

I

Ignition system:-
CDI unit 116
coil testing 116
electronic ignition system:
principles 114
testing and maintenance 114
exciter coils 116
fault diagnosis 11, 12, 13
ignition coil 116
ignition timing 30, 117
pulser coil 116

spark plugs:
checking the gap settings 117
operating conditions – colour chart 115
specifications 112
switch 170
wiring, connectors and switches 114

K

Kickstart:-
reassembly 73
removal 45

L

Lamps 162-165
Legal obligations 8, 24, 134, 157, 162
Lubrication:-
castrol lubricants 32
chain 29
changing transmission oil 28
checking transmission oil 25
control cables 23
controls, cables and pivots 25
final drive chain 156
general lubrication 29
reed valve induction system 110
rear chain 23
the engine lubrication system 105
topping up the oil tank 21

M

Maintenance – routine 20-31

O

Oil pump:-
bleeding 109
checking 26, 109
removing and replacing 107
Oil seals – examination and renovation 58

P

Pedal – rear brake 134
Pistons and rings 40, 59, 79
Petrol tank and cap 96

R

Reed valve induction system 110
Rear brake pedal 134
Rear chain 23, 29, 156
Rear suspension unit 133
Rear wheel:-
cush drive assembly:
RD350 LC 153
RD250 LC 153
sprocket:
RD350 LC 153
RD250 LC 153
Regulator/rectifier unit 160
Rings and pistons 40, 59, 79
Routine maintenance 20 – 31

S

Safety precautions 8
Spark plugs:-
   checking and resetting gaps 117
   operating conditions – 115
Specifications:-
   clutch 34
   cooling system 87
   electrical system 158
   engine 33, 34
   frame and forks 118
   fuel system 95
   gearbox 34
   ignition 112
   lubrication 96
   wheels, brakes and tyres 136
Speedometer and tachometer drives:-
   examination and maintenance 134
   location 134
Statutory requirements 8. 24, 134, 157, 162
Steering:-
   head assembly 121
   head bearings 28, 126
   lock 126
Suspension unit – rear 28, 133
Switches 169, 170

T

Tools 9
Torque wrench settings 34
Tyres:-
   pressures 22
   removal and replacement – 151

V

Valves:-
   reed induction system 110
   tyre 157

W

Weights and dimensions 6
Wheels:-
   bearings 149
   front 137, 143
   rear 146, 154
Wiring diagram 172
Working conditions 9